Operator Theory: Advances and
Applications
Vol. 136

Editor:
I. Gohberg

Editorial Office:
School of Mathematical
Sciences
Tel Aviv University
Ramat Aviv, Israel

Wavelet Transforms and Localization Operators

M. W. Wong

Springer Basel AG

Author:

M.W. Wong
Department of Mathematics and Statistics
York University
4700 Keele Street
Toronto, Ontario M3J IP3
Canada
e-mail: mwwong@pascal.math.yorku.ca

2000 Mathematics Subject Classification 47-02, 47G10, 47G30; 22A10, 42C40, 81S40, 94A12

A CIP catalogue record for this book is available from the
Library of Congress, Washington D.C., USA

Deutsche Bibliothek Cataloging-in-Publication Data

Wong, Man-Wah:
Wavelet transforms and localization operators / M. W. Wong. - Basel ;
Boston ; Berlin : Birkhäuser, 2002
  (Operator theory ; Vol. 136)
  ISBN 978-3-0348-9478-4   ISBN 978-3-0348-8217-0 (eBook)
  DOI 10.1007/978-3-0348-8217-0

© 2002 Springer Basel AG
Originally published by Birkhäuser Verlag in 2002
Softcover reprint of the hardcover 1st edition 2002
Printed on acid-free paper produced from chlorine-free pulp. TCF ∞
Cover design: Heinz Hiltbrunner, Basel

ISBN 978-3-0348-9478-4

Contents

Preface

This book is based on lectures given at the Global Analysis Research Center (GARC) of Seoul National University in 1999 and at Peking University in 1999 and 2000. Preliminary versions of the book have been used for various topics courses in analysis for graduate students at York University.

We study in this book wavelet transforms and localization operators in the context of infinite-dimensional and square-integrable representations of locally compact and Hausdorff groups. The wavelet transforms studied in this book, which include the ones that come from the Weyl-Heisenberg group and the well-known affine group, are the building blocks of localization operators. The theme that dominates the book is the spectral theory of wavelet transforms and localization operators in the form of Schatten-von Neumann norm inequalities. Several chapters are also devoted to the product formulas for concrete localization operators such as Daubechies operators and wavelet multipliers.

This book is a natural sequel to the book on pseudo-differential operators [103] and the book on Weyl transforms [102] by the author. Indeed, localization operators on the Weyl-Heisenberg group are Weyl transforms, which are in fact pseudo-differential operators. Details on the perspective and the organization of the book are laid out in the first chapter.

This is a book on mathematics and is written for anyone who has taken basic graduate courses in measure theory and functional analysis. Some knowledge of group theory and general topology at the undergraduate level is also assumed. As such, the book is suitable for graduate students and mathematicians who are interested in operator theory and harmonic analysis.

1 Introduction

The study of wavelet transforms and localization operators undertaken in this book can best be motivated by the study of a class of pseudo-differential operators, which we now recall.

Let $x = (x_1, x_2, \ldots, x_n)$ and $y = (y_1, y_2, \ldots, y_n)$ be any two points in $\mathbb{R}^n$. The inner product $x \cdot y$ of x and y is defined by

$$x \cdot y = \sum_{j=1}^{n} x_j y_j$$

and the norm $|x|$ of x is defined by

$$|x| = \left(\sum_{j=1}^{n} x_j^2 \right)^{\frac{1}{2}}.$$

For $j = 1, 2, \ldots$, we denote $\frac{\partial}{\partial x_j}$ by ∂_j and we define the partial differential operator D_j by

$$D_j = -i\partial_j,$$

where i is the complex number such that $i^2 = -1$.

Let α be a multi-index, i.e., $\alpha = (\alpha_1, \alpha_2, \ldots, \alpha_n)$, where $\alpha_1, \alpha_2, \ldots, \alpha_n$ are nonnegative integers. Then we define $|\alpha|$, ∂^α and D^α by

$$|\alpha| = \sum_{j=1}^{n} \alpha_j, \qquad \partial^\alpha = \partial_1^{\alpha_1} \partial_2^{\alpha_2} \ldots \partial_n^{\alpha_n} \qquad \text{and} \qquad D^\alpha = D_1^{\alpha_1} D_2^{\alpha_2} \ldots D_n^{\alpha_n}$$

respectively. We call $|\alpha|$ the length of the multi-index α. For all x in $\mathbb{R}^n$, we also define x^α by

$$x^\alpha = x_1^{\alpha_1} x_2^{\alpha_2} \ldots x_n^{\alpha_n}.$$

For $m \in \mathbb{R}$, we let S^m be the set of all complex-valued functions σ in $C^\infty(\mathbb{R}^n \times \mathbb{R}^n)$ such that for all multi-indices α and β, there exists a positive constant $C_{\alpha,\beta}$ for which

$$|(D_x^\alpha D_\xi^\beta \sigma)(x, \xi)| \le C_{\alpha,\beta}(1 + |\xi|)^{m-\beta}, \quad x, \xi \in \mathbb{R}^n.$$

Let $\sigma \in S^m$. Then we call σ a symbol of order m and we define a linear operator T_σ on the Schwartz space $\mathcal{S}$ by

$$(T_\sigma \varphi)(x) = (2\pi)^{-\frac{n}{2}} \int_{\mathbb{R}^n} e^{ix \cdot \xi} \sigma(x, \xi) \hat{\varphi}(\xi) d\xi, \quad \varphi \in \mathcal{S},$$

where the Schwartz space $\mathcal{S}$ is defined to be the set of all complex-valued functions φ in $C^\infty(\mathbb{R}^n)$ such that

$$\sup_{x \in \mathbb{R}^n} |x^\gamma (D^\delta \varphi)(x)| < \infty$$

for all multi-indices γ and δ, and $\hat{\varphi}$ is the Fourier transform of φ defined by

$$\hat{\varphi}(\xi) = (2\pi)^{-\frac{n}{2}} \int_{\mathbb{R}^n} e^{-ix \cdot \xi} \varphi(x) dx, \quad \xi \in \mathbb{R}^n.$$

We call T_σ the pseudo-differential operator associated to the symbol σ. The most fundamental properties of pseudo-differential operators which are useful in the study of partial differential equations are listed in Theorems 1.1–1.4.

Theorem 1.1 *If $\sigma \in S^{m_1}$ and $\tau \in S^{m_2}$, then $T_\sigma T_\tau = T_\lambda$, where $\lambda \in S^{m_1+m_2}$ and*

$$\lambda \sim \sum_\mu \frac{(-i)^{|\mu|}}{\mu!} (\partial_\xi^\mu \sigma)(\partial_x^\mu \tau).$$

The asymptotic expansion $\lambda \sim \sum_\mu \frac{(-i)^{|\mu|}}{\mu!} (\partial_\xi^\mu \sigma)(\partial_x^\mu \tau)$ in Theorem 1.1 means that

$$\lambda - \sum_{|\mu| < N} \frac{(-i)^{|\mu|}}{\mu!} (\partial_\xi^\mu \sigma)(\partial_x^\mu \tau) \in S^{m_1+m_2-N}$$

for all positive integers N.

Theorem 1.2 *If $\sigma \in S^m$, then $T_\sigma^+ = T_\tau$, where $\tau \in S^m$ and*

$$\tau \sim \sum_\mu \frac{(-i)^{|\mu|}}{\mu!} \partial_x^\mu \partial_\xi^\mu \overline{\sigma}.$$

Here, T_σ^+ is the formal adjoint of T_σ.

In Theorem 1.2, the formal adjoint T_σ^+ of T_σ is defined by

$$(T_\sigma \varphi, \psi)_{L^2(\mathbb{R}^n)} = (\varphi, T_\sigma^+ \psi)_{L^2(\mathbb{R}^n)}$$

for all φ and ψ in $\mathcal{S}$, where $(\ ,\ )_{L^2(\mathbb{R}^n)}$ is the inner product in $L^2(\mathbb{R}^n)$. The asymptotic expansion $\tau \sim \sum_\mu \frac{(-i)^{|\mu|}}{\mu!} \partial_x^\mu \partial_\xi^\mu \overline{\sigma}$ means that

$$\tau - \sum_{|\mu| < N} \frac{(-i)^{|\mu|}}{\mu!} \partial_x^\mu \partial_\xi^\mu \overline{\sigma} \in S^{m-N}$$

for all positive integers N.

Before we state Theorem 1.3, let us recall that a symbol σ in S^m is said to be elliptic if there exist positive constants C and R such that

$$|\sigma(x,\xi)| \geq C(1 + |\xi|)^m, \quad |\xi| \geq R.$$

Of course, a pseudo-differential operator T_σ is said to be elliptic if σ is elliptic.

Theorem 1.3 *Let $\sigma \in S^m$. Then there exists a symbol τ in S^{-m} such that $T_\tau T_\sigma = I + R$ and $T_\sigma T_\tau = I + S$, where R and S are pseudo-differential operators with symbols in $\bigcap_{k \in \mathbb{R}} S^k$ if and only if σ is elliptic.*

Pseudo-differential operators with symbols in $\bigcap_{k \in \mathbb{R}} S^k$ are considered to be negligible in the regularity theory of partial differential equations. Thus, Theorem 1.3 says that elliptic pseudo-differential operators have approximate inverses. Approximate inverses are also known as parametrices.

Theorem 1.4 *If $\sigma \in S^0$, then T_σ, initially defined on S, can be uniquely extended to a bounded linear operator from $L^2(\mathbb{R}^n)$ into $L^2(\mathbb{R}^n)$.*

Theorem 1.4 allows us to study pseudo-differential operators as bounded linear operators on $L^2(\mathbb{R}^n)$.

The theory of pseudo-differential operators that we have described can be found in Hörmander [47], Kumano-go [58], Saint Raymond [75], Shubin [81], Stein [87], Taylor [93], Treves [96], Wong [103] and others.

The class of pseudo-differential operators that we have introduced arise naturally in quantum physics. To wit, let us begin with a very simple sketch of the foundations of physics. It is well known that among the fundamental objects of study are phase spaces and observables. In classical mechanics, the phase space is $\mathbb{R}^n \times \mathbb{R}^n$ and the observables are given by real-valued functions on $\mathbb{R}^n \times \mathbb{R}^n$. In quantum mechanics, the phase space is $L^2(\mathbb{R}^n)$ and the observables are self-adjoint operators on $L^2(\mathbb{R}^n)$. The mechanism that allows us to pass from classical mechanics to quantum mechanics is known as quantization. The rules of quantization due to von Neumann [97] consist of replacing the position x_j in the jth coordinate by the multiplication operator by x_j, and substituting the momentum ξ_j in the jth coordinate by the partial differential operator D_j. We make this more transparent by means of an example in relativistic quantum mechanics. The book [1] by Aitchison and the book [6] by Bjorken and Drell are good references in relativistic quantum mechanics.

Example 1.5 Consider a relativistic particle of mass m moving on $\mathbb{R}$ under the influence of a potential $V(x)$. Then the momentum ξ and the kinetic energy E_k are given, respectively, by

$$\xi = \frac{mv}{\sqrt{1 - \frac{v^2}{c^2}}} \qquad \text{and} \qquad E_k = \frac{mc^2}{\sqrt{1 - \frac{v^2}{c^2}}},$$

where v is the velocity of the particle and c is the velocity of light. Following the standard practice of mathematicians, we let $m = c = 1$. Then

$$\begin{cases} \xi = \frac{v}{\sqrt{1 - v^2}}, \\ E_k = \frac{1}{\sqrt{1 - v^2}}. \end{cases}$$

So,

$$E_k^2 = 1 + \xi^2 \Rightarrow E_k = \sqrt{1 + \xi^2},$$

and hence by von Neumann's rules of quantization, the total energy E of the relativistic particle is given by

$$E = \begin{cases} \sqrt{1 + \xi^2} + V(x), & \text{Classical,} \\ \sqrt{1 - \frac{d^2}{dx^2}} + V(x), & \text{Quantum.} \end{cases}$$

If we let σ be the function on $\mathbb{R} \times \mathbb{R}$ defined by

$$\sigma(x, \xi) = \sqrt{1 + \xi^2} + V(x), \quad x, \xi \in \mathbb{R},$$

then the total energy E of the relativistic particle in quantum mechanics is given by

$$E = \sigma(x, D),$$

where $D = -i\frac{d}{dx}$.

What is the linear operator $\sigma(x, D)$ that arises in quantizing the relativistic energy in classical mechanics in accordance with von Neumann's rules of quantization? A plausible answer is provided by the class of pseudo-differential operators that we have just introduced. We can try to define $\sigma(x, D)$ as T_σ. Is $\sigma \to T_\sigma$ a quantization? To answer this question, we need to recall that a requirement for a correct quantization is that the linear operator T_σ corresponding to a real-valued symbol σ should be self-adjoint and hence at least formally self-adjoint. However, by Theorem 1.2, we see that for a real-valued symbol σ, $T_\sigma^+ \neq T_\sigma$ in general. So, $\sigma \to T_\sigma$ is not a quantization.

In order to arrive at a correct quantization, let us recall that for all x in $\mathbb{R}^n$,

$$\begin{aligned} (T_\sigma \varphi)(x) &= (2\pi)^{-\frac{n}{2}} \int_{\mathbb{R}^n} e^{ix \cdot \xi} \sigma(x, \xi) \hat{\varphi}(\xi) d\xi \\ &= (2\pi)^{-n} \int_{\mathbb{R}^n} \int_{\mathbb{R}^n} e^{i(x-y)\cdot\xi} \sigma(x, \xi) \varphi(y) dy\, d\xi, \end{aligned}$$

where the last integral is to be understood as an iterated integral in which the integration with respect to y has to be performed first. In fact, we can associate to σ another linear operator W_σ on $\mathcal{S}$ defined by

$$(W_\sigma \varphi)(x) = (2\pi)^{-n} \int_{\mathbb{R}^n} \int_{\mathbb{R}^n} e^{i(x-y)\cdot\xi} \sigma\left(\frac{x+y}{2}, \xi\right) \varphi(y) dy\, d\xi \qquad (1.1)$$

for all φ in $\mathcal{S}$. This formula can be traced back to the work [100] by Weyl and hence we call W_σ the Weyl transform associated to the symbol σ. In fact, we have the following connection between Weyl transforms and pseudo-differential operators.

Theorem 1.6 *Let $\sigma \in S^m$. Then there exists a symbol τ in S^m such that $T_\sigma = W_\tau$ and there exists a symbol κ in S^m such that $W_\sigma = T_\kappa$.*

Thus, there is a one-to-one correspondence between the set of pseudo-differential operators and the set of Weyl transforms.

One of the most important properties of Weyl transforms is the following theorem.

Theorem 1.7 *Let $\sigma \in S^m$. Then $W_\sigma^+ = W_{\bar{\sigma}}$, where W_σ^+ is the formal adjoint of W_σ.*

An immediate consequence of Theorem 1.7 is the following fact.

Corollary 1.8 *Let $\sigma \in S^m$ be real-valued. Then $W_\sigma^+ = W_\sigma$.*

In view of Corollary 1.8, we can conclude that the Weyl transform W_σ associated to a real-valued symbol is formally self-adjoint. Hence $\sigma \to W_\sigma$ is a good candidate for quantization. That this is indeed the correct quantization is explained in the book [102] by Wong.

Now, let $\sigma \in S^0$ be real-valued. Then, by Theorems 1.4 and 1.6, W_σ can be extended uniquely to a bounded linear operator, again denoted by W_σ, from $L^2(\mathbb{R}^n)$ into $L^2(\mathbb{R}^n)$. By Corollary 1.8, it can be proved that $W_\sigma : L^2(\mathbb{R}^n) \to L^2(\mathbb{R}^n)$ is self-adjoint and hence a bona fide observable in quantum mechanics. It is a doctrine in quantum mechanics that the numerical values of the measurements of the observable $W_\sigma : L^2(\mathbb{R}^n) \to L^2(\mathbb{R}^n)$ are precisely provided by the spectrum $\Sigma(W_\sigma)$. Thus, the spectral analysis of the linear operator $W_\sigma : L^2(\mathbb{R}^n) \to L^2(\mathbb{R}^n)$ or, equivalently, the study of the spectrum $\Sigma(W_\sigma)$, is of paramount importance in mathematics. Unfortunately, this is often a very difficult problem and very little is known about the spectrum $\Sigma(W_\sigma)$ of a Weyl transform associated to a symbol in S^0. Results on the spectra $\Sigma(W_\sigma)$ of Weyl transforms associated to symbols in $L^1(\mathbb{R}^n \times \mathbb{R}^n) \cup L^2(\mathbb{R}^n \times \mathbb{R}^n)$ or modulation spaces satisfying appropriate conditions do exist and can be found in, e.g., the papers [16, 17] by Du and Wong, [42] by Heil, Ramanathan and Topiwala, and [70] by Ramanathan and Topiwala. The connection between modulation spaces and pseudo-differential operators can also be found in [33] by Gröchenig, [34] by Gröchenig and Heil, and [92] by Tachizawa.

We can now look at some recent developments in wavelet analysis that shed some light on the spectral analysis of pseudo-differential operators and Weyl transforms.

Let U be the upper half plane given by

$$U = \{(b, a) : b \in \mathbb{R}, a > 0\}.$$

Then we define the binary operation $\cdot$ on U by

$$(b_1, a_1) \cdot (b_2, a_2) = (b_1 + a_1 b_2, a_1 a_2)$$

for all (b_1, a_1) and (b_2, a_2) in U. With respect to the binary operation $\cdot$, U is a non-abelian group in which $(0, 1)$ is the identity element and the inverse element of (b, a) is $\left(-\frac{b}{a}, \frac{1}{a}\right)$ for all (b, a) in U. In fact, U is a locally compact and Hausdorff group on which the left and right Haar measures are given by

$$d\mu = \frac{db\, da}{a^2} \quad \text{and} \quad d\nu = \frac{db\, da}{a}$$

respectively. The group U is called the affine group and is non-unimodular.

Let $H_+^2(\mathbb{R})$ be the subspace of $L^2(\mathbb{R})$ defined by

$$H_+^2(\mathbb{R}) = \{f \in L^2(\mathbb{R}) : \operatorname{supp}(\hat{f}) \subseteq [0, \infty)\},$$

where $\operatorname{supp}(\hat{f})$ is the support of $\hat{f}$. The function $\hat{f}$ is the Fourier transform of f defined by

$$\hat{f}(\xi) = \lim_{R \to \infty} \frac{1}{\sqrt{2\pi}} \int_{-\infty}^{\infty} e^{-ix\xi} \chi_R(x) f(x) dx, \quad \xi \in \mathbb{R},$$

where χ_R is the characteristic function on the interval $[-R, R]$. We can also define $H_-^2(\mathbb{R})$ to be the subspace of $L^2(\mathbb{R})$ by

$$H_-^2(\mathbb{R}) = \{f \in L^2(\mathbb{R}) : \operatorname{supp}(\hat{f}) \subseteq (-\infty, 0]\}.$$

$H_+^2(\mathbb{R})$ and $H_-^2(\mathbb{R})$ are known as the Hardy space and the conjugate Hardy space respectively. They can be shown to be closed subspaces of $L^2(\mathbb{R})$.

Let $U(H_+^2(\mathbb{R}))$ be the group of all unitary operators on $H_+^2(\mathbb{R})$ and let $\pi : U \to U(H_+^2(\mathbb{R}))$ be the mapping defined by

$$(\pi(b, a)f)(x) = \frac{1}{\sqrt{a}} f\left(\frac{x-b}{a}\right), \quad x \in \mathbb{R}. \tag{1.2}$$

Then it can be proved that $\pi : U \to U(H_+^2(\mathbb{R}))$ is an irreducible and unitary representation of U on $H_+^2(\mathbb{R})$. In fact, it can be proved that $\pi : U \to U(H_+^2(\mathbb{R}))$ is a square-integrable representation of U on $H_+^2(\mathbb{R})$ in the sense that there exists a function φ in $H_+^2(\mathbb{R})$ such that $\|\varphi\|_{L^2(\mathbb{R})} = 1$ and

$$\int_U |(\varphi, \pi(b, a)\varphi)_{L^2(\mathbb{R})}|^2 d\mu(b, a) < \infty.$$

Such a function φ is known as an admissible wavelet for the square-integrable representation $\pi : U \to U(H_+^2(\mathbb{R}))$ of U on $H_+^2(\mathbb{R})$. It can be proved that the admissibility condition holds if and only if

$$\int_0^\infty \frac{|\hat{\varphi}(\xi)|^2}{\xi} d\xi < \infty.$$

The above-mentioned facts on the affine group and the square-integrable representation $\pi : U \to U(H_+^2(\mathbb{R}))$ can be found in the paper [43] by Heil and Walnut and the book [104] by Wong.

Let φ be an admissible wavelet for the square-integrable representation $\pi :$ $U \to U(H_+^2(\mathbb{R}))$ of U on $H_+^2(\mathbb{R})$. Then the wavelet transform $A_\varphi : H_+^2(\mathbb{R}) \to L^2(U)$ associated to the admissible wavelet φ is defined by

$$(A_\varphi f)(b, a) = \frac{1}{\sqrt{c_\varphi}}(f, \pi(b, a)\varphi)_{L^2(\mathbb{R})}, \quad (b, a) \in U,$$

where

$$c_\varphi = \int_U |(\varphi, \pi(b, a)\varphi)_{L^2(\mathbb{R})}|^2 d\mu(b, a).$$

The wavelet transform just introduced is the one that has been studied most extensively in mathematics, statistics, science and engineering. See, for instance, Blatter [7], Chan [10], Chui [11], Daubechies [13], Gasquet and Witomsky [29], Gröchenig [33], Hernández and Weiss [44], Holschneider [46], Kaiser [49], Krantz [57], Meyer [63], Pinsky [67], Rao and Bopardikar [71], Strang and Nguyen [89], Strichartz [90], Walker [98, 99], Wojtaszczyk [101] and Young [107] in this connection. However, we prefer to look at the wavelet transform in the context of square-integrable representations of locally compact and Hausdorff groups on infinite-dimensional, separable and complex Hilbert spaces. The group representation that underscores this most widely used and most popular wavelet transform is the representation $\pi : U \to U(H_+^2(\mathbb{R}))$ of the affine group U on the Hardy space $H_+^2(\mathbb{R})$ defined by (1.2). Looking at wavelet transforms in this perspective is a fairly recent venture and has been studied in Ali, Antoine and Gazeau [2], Ali, Antoine, Gazeau and Mueller [3], Du and Wong [19], Du, Wong and Zhang [23], J. He [37], J. He and Liu [38], Z. He [39], Jiang and Peng [48], Kawazoe [51, 52, 53], Liu [61], Liu and Peng [62], Stark [86] and Wong [104, 105]. Of particular interest in the study of pseudo-differential operators and Weyl transforms is the case when the group representation is the Schrödinger representation of the Weyl-Heisenberg group $(WH)^n$ on $L^2(\mathbb{R}^n)$, which we now describe.

Let $\mathbb{R}^n \times \mathbb{R}^n = \{(q, p) : q, p \in \mathbb{R}^n\}$ and let $\mathbb{Z}$ be the set of all integers. Let $(WH)^n = \mathbb{R}^n \times \mathbb{R}^n \times \mathbb{R}/2\pi\mathbb{Z}$. Then we define the binary operation $\cdot$ on $(WH)^n$ by

$$(q_1, p_1, t_1) \cdot (q_2, p_2, t_2) = (q_1 + q_2, p_1 + p_2, t_1 + t_2 + q_1 \cdot p_2)$$

for all points (q_1, p_1, t_1) and (q_2, p_2, t_2) in $(WH)^n$, where t_1, t_2 and $t_1 + t_2 + q_1 \cdot p_2$ are cosets in the quotient group $\mathbb{R}/2\pi\mathbb{Z}$ in which the group law is addition modulo 2π. With respect to the binary operation $\cdot$, $(WH)^n$ is a non-abelian group in which $(0, 0, 0)$ is the identity element and the inverse element of (q, p, t) is $(-q, -p, -t + q \cdot p)$ for all (q, p, t) in $(WH)^n$. If we identify $\mathbb{R}/2\pi\mathbb{Z}$ with the interval $[0, 2\pi]$, then $(WH)^n$ can be identified with $\mathbb{R}^n \times \mathbb{R}^n \times [0, 2\pi]$. It is a locally compact and

Hausdorff group on which the left and right Haar measure is the Lebesgue measure $dq\,dp\,dt$. $(WH)^n$ is known as the Weyl-Heisenberg group and it is unimodular.

Let $U(L^2(\mathbb{R}^n))$ be the group of all unitary operators on $L^2(\mathbb{R}^n)$ and let $\pi : (WH)^n \to U(L^2(\mathbb{R}^n))$ be the mapping defined by

$$(\pi(q,p,t)f)(x) = e^{i(p\cdot x - q\cdot p + t)}f(x-q), \quad x \in \mathbb{R}^n,$$

for all (q,p,t) in $(WH)^n$ and all f in $L^2(\mathbb{R}^n)$. Then $\pi : (WH)^n \to U(L^2(\mathbb{R}^n))$ is an irreducible and unitary representation of $(WH)^n$ on $L^2(\mathbb{R}^n)$. In fact, it is a square-integrable representation of $(WH)^n$ on $L^2(\mathbb{R}^n)$ in the sense that every function φ in $L^2(\mathbb{R}^n)$ with $\|\varphi\|_{L^2(\mathbb{R}^n)} = 1$ satisfies the admissibility condition that

$$\int_{(WH)^n} |(\varphi, \pi(q,p,t)\varphi)_{L^2(\mathbb{R}^n)}|^2 dq\,dp\,dt < \infty.$$

Thus, every function φ in $L^2(\mathbb{R}^n)$ with $\|\varphi\|_{L_2(\mathbb{R}^n)} = 1$ is an admissible wavelet for the square-integrable representation $\pi : (WH)^n \to U(L^2(\mathbb{R}^n))$ of $(WH)^n$ on $L^2(\mathbb{R}^n)$. The representation $\pi : (WH)^n \to U(L^2(\mathbb{R}^n))$ is often called the Schrödinger representation of $(WH)^n$ on $L^2(\mathbb{R}^n)$. The paper [43] and the book [104] contain the basic facts on the Weyl-Heisenberg group and the Schrödinger representation.

Let $\varphi \in L^2(\mathbb{R}^n)$ be such that $\|\varphi\|_{L^2(\mathbb{R}^n)} = 1$. Then the wavelet transform $A_\varphi : L^2(\mathbb{R}^n) \to L^2((WH)^n)$ associated to the admissible wavelet φ is defined by

$$(A_\varphi f)(q,p,t) = \frac{1}{\sqrt{c_\varphi}}(f, \pi(q,p,t)\varphi)_{L^2(\mathbb{R}^n)}, \quad (q,p,t) \in (WH)^n,$$

where

$$c_\varphi = \int_{(WH)^n} |(\varphi, \pi(q,p,t)\varphi)_{L^2(\mathbb{R}^n)}|^2 dq\,dp\,dt.$$

In fact, $c_\varphi = (2\pi)^{n+1}$.

In the paper [12] by Daubechies, a class of bounded linear operators $D_{F,\varphi} : L^2(\mathbb{R}^n) \to L^2(\mathbb{R}^n)$ associated to F in $L^1(\mathbb{R}^n \times \mathbb{R}^n)$ and φ in $L^2(\mathbb{R}^n)$ with $\|\varphi\|_{L^2(\mathbb{R}^n)} = 1$ is studied in the context of signal analysis. In fact,

$$(D_{F,\varphi}u,v)_{L^2(\mathbb{R}^n)} = (2\pi)^{-n}\int_{\mathbb{R}^n}\int_{\mathbb{R}^n} F(q,p)(u,\varphi_{q,p})_{L^2(\mathbb{R}^n)}(\varphi_{q,p},v)_{L^2(\mathbb{R}^n)}dq\,dp$$

for all u and v in $L^2(\mathbb{R}^n)$, where

$$\varphi_{q,p}(x) = e^{ip\cdot x}\varphi(x-q), \quad x \in \mathbb{R}^n,$$

for all q and p in $\mathbb{R}^n$. We can prove that

$$(D_{F,\varphi}u,v)_{L^2(\mathbb{R}^n)}$$
$$= \frac{1}{\sqrt{c_\varphi}}\int_{(WH)^n} F(q,p)(u,\pi(q,p,t)\varphi)_{L^2(\mathbb{R}^n)}(\pi(q,p,t)\varphi,v)_{L^2(\mathbb{R}^n)}dq\,dp\,dt$$

for all u and v in $L^2(\mathbb{R}^n)$. Thus, the linear operator $D_{F,\varphi} : L^2(\mathbb{R}^n) \to L^2(\mathbb{R}^n)$, which is called the Daubechies operator in [20, 21], is the same as the localization operator $L_{F,\varphi} : L^2(\mathbb{R}^n) \to L^2(\mathbb{R}^n)$ associated to the symbol F and the admissible wavelet φ for the Schrödinger representation of the Weyl-Heisenberg group $(WH)^n$ on $L^2(\mathbb{R}^n)$. A Daubechies operator associated to an admissible wavelet is like a windowed Fourier transform used by Gabor [28] in time-frequency analysis. The admissible wavelet plays the same role in the Daubechies operator that the window plays in the windowed Fourier transform. A symbolic calculus for the product of two such localization operators is given in [18].

Daubechies operators are also the same as pseudo-differential operators with anti-Wick symbols studied by Boggiatto, Buzano and Rodino [8], and Shubin [81], among others, in the context of quantization.

It is a result, *i.e.*, Theorem 17.1 in the book [102] by Wong, that if φ is the function on $\mathbb{R}^n$ defined by

$$\varphi(x) = \pi^{-\frac{n}{4}} e^{-\frac{|x|^2}{2}}, \quad x \in \mathbb{R}^n,$$

then the Daubechies operator $D_{F,\varphi} : L^2(\mathbb{R}^n) \to L^2(\mathbb{R}^n)$ is the same as the Weyl transform $W_{F*\Lambda} : L^2(\mathbb{R}^n) \to L^2(\mathbb{R}^n)$, where

$$\Lambda(x,\xi) = \pi^{-n} e^{-(|x|^2+|\xi|^2)}, \quad x, \xi \in \mathbb{R}^n,$$

and $F * \Lambda$ is the convolution of F and Λ given by

$$(F * \Lambda)(x,\xi) = \int_{\mathbb{R}^n} \int_{\mathbb{R}^n} F(x-y, \xi-\eta)\Lambda(y,\eta)dy\,d\eta, \quad x, \xi \in \mathbb{R}^n.$$

Motivated by Daubechies operators, which are localization operators on the Weyl-Heisenberg group $(WH)^n$ equipped with the Schrödinger representation of $(WH)^n$ on $L^2(\mathbb{R}^n)$, we give a systematic study of wavelet transforms and localization operators on locally compact and Hausdorff groups G in this book. These wavelet transforms and localization operators are based on coherent states parametrized by elements in the group G and admissible wavelets defined in terms of the coherent states. Daubechies operators turn out to be just localization operators on the Weyl-Heisenberg group defined in terms of the coherent states first envisaged in the 1926 paper [78] by Schrödinger. The books [56] by Klauder and Skagerstam and [66] by Perelomov are definitive accounts on coherent states.

In another direction, guided by the Landau-Pollak-Slepian operator in signal analysis, a theory of wavelet multipliers has been initiated in the paper [41] by He and Wong, developed in the paper [22] by Du and Wong, and detailed in the book [104] by Wong. Wavelet multipliers are localization operators on the additive group $\mathbb{R}^n$ defined in terms of coherent states parametrized by points in $\mathbb{R}^n$.

In the case of localization operators on a locally compact and Hausdorff group G endowed with a left Haar measure, the coherent states originate from a unitary representation of G on an infinite-dimensional, separable and complex Hilbert space. For wavelet multipliers, the coherent states stem from the unitary representation of $\mathbb{R}^n$ on $L^2(\mathbb{R}^n)$ given by modulation. Suggested by the book [2] of Ali, Antoine and Gazeau, and the paper [3] of Ali, Antoine, Gazeau and Mueller, the Daubechies operators, localization operators on locally compact and Hausdorff groups and wavelet multipliers can be looked at as localization operators on homogeneous spaces. The book [2] contains an extensive list of references on coherent states parametrized by points in a homogeneous space.

The aim of this book is to give a compact account of the Schatten-von Neumann property of localization operators and wavelet multipliers, which include the very important class of Weyl transforms dubbed as Daubechies operators. Basic to the construction of localization operators and wavelet multipliers are the wavelet transforms on locally compact and Hausdorff groups, which are also studied in this book as objects of interest in their own right. We also give a symbolic calculus for Daubechies operators and two symbolic calculi for wavelet multipliers. These symbolic calculi are rudimentary in the sense that only the product of two Daubechies operators and the product of two wavelet multipliers are considered.

The genesis of the book is as follows. Chapters 2–5 are devoted to background materials which are fundamental in an understanding of the contents in this book. This book really begins with Chapter 6 in which an exposition of the theory of square-integrable representations is given. The exposition is based on the excellent paper [36] by Grossmann, Morlet and Paul. Earlier contributions to this subject include the works of Carey [9], Dixmier [14], and Duflo and Moore [24]. Admissible wavelets and wavelet transforms, which inevitably arise in the study of square-integrable representations, are studied in Chapters 7–11 as objects of interest in their own right. Localization operators associated to admissible wavelets for square-integrable representations of locally compact and Hausdorff groups on infinite-dimensional, separable and complex Hilbert spaces are introduced in Chapter 12. The Schatten-von Neumann property in general, the trace class property and the Hilbert-Schmidt theory in particular, for these localization operators are developed in Chapters 13–16. We specialize to the Weyl-Heisenberg group in Chapter 17, the affine group in Chapter 18, wavelet multipliers in Chapter 19 and the Landau-Pollak-Slepian operator in Chapter 20. These four canonical examples of localization operators are the main impetus for the abstract theory developed in the book and their importance can hardly be over-emphasized. Product formulas for wavelet multipliers are given in Chapter 21. The more important formula for the product of two localization operators is derived in Chapter 22 using a product formula of Grossmann, Loupias and Stein [35] for Weyl transforms. To this end, a new twisted convolution for functions in $L^2(\mathbb{R}^n \times \mathbb{R}^n)$ is introduced, and in Chapter 23 we tackle the problem of finding subspaces M of $L^2(\mathbb{R}^n \times \mathbb{R}^n)$ such that the new twisted convolution is a binary operation on M. In the last three chapters, we introduce localization operators on homogeneous spaces and show that they can be used to unify the various localization operators studied in the book into a single theory.

2   Schatten-von Neumann Classes

We devote this chapter to a brief and fairly self-contained study on the Schatten-von Neumann classes. The topics are selected and organized in such a way that we can use them easily later in this book. Some proofs are omitted whenever they are easily available in the literature. More comprehensive accounts on the Schatten-von Neumann classes can be found in Dunford and Schwartz [25], Gohberg, Goldberg and Krupnik [31], Reed and Simon [72], Simon [82] and Zhu [108]. Recommended references for basic functional analysis that we use in this book are Douglas [15], Gohberg and Goldberg [30], Reed and Simon [72], Schechter [76] and Young [106].

Let X be a separable and complex Hilbert space in which the inner product and the norm are denoted by $(\,,\,)$ and $\|\;\|$ respectively. Let $A : X \to X$ be a bounded linear operator. We define $|A| : X \to X$ by $|A| = (A^*A)^{\frac{1}{2}}$, where A^* is the adjoint of A. We call $|A| : X \to X$ the absolute value of $A : X \to X$. A bounded linear operator $V : X \to X$ is said to be a partial isometry if V is an isometry when restricted to the orthogonal complement $N(V)^{\perp}$ of the null space $N(V)$ of V. The starting point is the following well-known theorem, which gives the polar form of a bounded linear operator on a separable and complex Hilbert space. A proof can be found on pages 88 and 89 of the book [15] by Douglas or on pages 197 and 198 of the book [72] by Reed and Simon.

Theorem 2.1 *Let $A : X \to X$ be a bounded linear operator. Then there exists a partial isometry $V : X \to X$ such that $A = V|A|$. The partial isometry $V : X \to X$ is uniquely determined by the condition $N(V) = N(A)$. Furthermore, the range of $V : X \to X$ is equal to the closure of the range of $A : X \to X$.*

Let $A : X \to X$ be a compact operator. Then the linear operator $|A| : X \to X$ is positive and compact. Let $\{\varphi_k : k = 1, 2, \ldots\}$ be an orthonormal basis for X consisting of eigenvectors of $|A| : X \to X$, and let $s_k(A)$ be the eigenvalue of $|A| : X \to X$ corresponding to the eigenvector φ_k, $k = 1, 2, \ldots$. We call $s_k(A)$, $k = 1, 2, \ldots$, the singular values of $A : X \to X$. A basic result that we need is the following canonical form for compact operators.

Theorem 2.2 *Let $A : X \to X$ be a compact operator. Then we can find an orthonormal basis $\{u_k : k = 1, 2, \ldots\}$ for $N(A)^{\perp}$ consisting of eigenvectors of $|A| : X \to X$ and an orthonormal set $\{v_k : k = 1, 2, \ldots\}$ in X such that*

$$A = \sum_{k=1}^{\infty} s_k(A)(\cdot, u_k)v_k,$$

where $s_k(A)$, $k = 1, 2, \ldots$, are the positive singular values of $A : X \to X$ and the series converges to A strongly.

Proof. By Theorem 2.1, we can write $A = V|A|$, where $V : X \to X$ is the partial isometry uniquely determined by the condition $N(V) = N(A)$. We write $X = N(V)^{\perp} \oplus N(V)$. Moreover,

$$x \in N(V) = N(A) \Rightarrow V|A|x = Ax = 0 \Rightarrow |A|x \in N(V)$$

and

$$x \in N(V)^{\perp}, y \in N(V) \Rightarrow (|A|x, y) = (x, |A|y) = 0.$$

Thus, $|A|x \in N(V)^{\perp}$. Therefore $N(V)^{\perp}$ and $N(V)$ are invariant subspaces of X with respect to $|A| : X \to X$. Thus, we can pick an orthonormal basis $\{u_k : k = 1, 2, \ldots\}$ for $N(V)^{\perp}$ and an orthonormal basis $\{w_k : k = 1, 2, \ldots\}$ for $N(V)$ consisting of eigenvectors of $|A| : X \to X$. Putting the orthonormal bases for $N(V)^{\perp}$ and for $N(V)$ together, we get an orthonormal basis for X. For $k = 1, 2, \ldots$, let $s_k(A)$ be the eigenvalue of $|A| : X \to X$ corresponding to u_k and let t_k be the eigenvalue of $|A| : X \to X$ corresponding to w_k. Then

$$|A|w_k = t_k w_k \Rightarrow 0 = A^* A w_k = |A|^2 w_k = t_k^2 w_k \Rightarrow t_k = 0$$

for $k = 1, 2, \ldots$. Now, the spectral theorem gives

$$|A| = \sum_{k=1}^{\infty} s_k(A)(\cdot, u_k) u_k,$$

where the series converges to A strongly. Hence

$$A = \sum_{k=1}^{\infty} s_k(A)(\cdot, u_k) V u_k,$$

where the series converges to A strongly. For $k = 1, 2, \ldots$, let $v_k = V u_k$. Since $u_k \in N(V)^{\perp}$, $k = 1, 2, \ldots$, and $V : X \to X$ is a partial isometry, it follows that $\{v_k : k = 1, 2, \ldots\}$ is an orthonormal set in X, and the proof is complete.   □

A compact operator $A : X \to X$ is said to be in the Schatten-von Neumann class S_p, $1 \leq p < \infty$, if

$$\sum_{k=1}^{\infty} (s_k(A))^p < \infty.$$

It can be shown that S_p, $1 \leq p < \infty$, is a complex Banach space in which the norm $\| \ \|_{S_p}$ is given by

$$\|A\|_{S_p} = \left\{ \sum_{k=1}^{\infty} (s_k(A))^p \right\}^{\frac{1}{p}}, \quad A \in S_p.$$

We let S_∞ be the C^*-algebra $B(X)$ of all bounded linear operators on X. Thus, $\| \ \|_{S_\infty} = \| \ \|_*$, where $\| \ \|_*$ denotes the norm in $B(X)$. It is obvious that $S_p \subseteq S_q$, $1 \le p \le q \le \infty$.

It is customary to call S_1 the trace class and S_2 the Hilbert-Schmidt class.

Proposition 2.3 *Let $A : X \to X$ be a positive operator. If there exists an orthonormal basis $\{\varphi_k : k = 1, 2, \ldots\}$ for X such that*

$$\sum_{k=1}^{\infty} (A\varphi_k, \varphi_k) < \infty,$$

then $A : X \to X$ is compact.

Proof. Let $B = \sqrt{A}$ and let $\{\varphi_k : k = 1, 2, \ldots\}$ be an orthonormal basis for X for which

$$\sum_{k=1}^{\infty} (A\varphi_k, \varphi_k) < \infty.$$

Then

$$\sum_{k=1}^{\infty} \|B\varphi_k\|^2 = \sum_{k=1}^{\infty} (A\varphi_k, \varphi_k) < \infty. \tag{2.1}$$

Now, there exist complex numbers b_{jk}, $j, k = 1, 2, \ldots$, such that

$$B\varphi_k = \sum_{j=1}^{\infty} b_{jk}\varphi_j, \tag{2.2}$$

where the convergence of the series is in X, and the Parseval's identity gives

$$\|B\varphi_k\|^2 = \sum_{j=1}^{\infty} |b_{jk}|^2$$

for $k = 1, 2, \ldots$. Thus, by (2.1) and (2.2),

$$\sum_{j,k=1}^{\infty} |b_{jk}|^2 = \sum_{k=1}^{\infty} \|B\varphi_k\|^2 < \infty. \tag{2.3}$$

For any positive integer N, let

$$b_{jk}^N = \begin{cases} b_{jk}, & 1 \le j, k \le N, \\ 0, & \text{otherwise.} \end{cases} \tag{2.4}$$

Let $B_N : X \to X$ be the bounded linear operator given by

$$B_N\varphi_k = \begin{cases} \sum_{j=1}^{N} b_{jk}^N \varphi_j, & 1 \le k \le N, \\ 0, & k > N. \end{cases} \tag{2.5}$$

Then $B_N : X \to X$ is a finite rank operator and for all x in X, we get, by Schwarz' inequality and Parseval's identity,

$$
\begin{aligned}
\|(B - B_N)x\| &= \left\| (B - B_N) \sum_{k=1}^{\infty} (x, \varphi_k)\varphi_k \right\| \\
&\leq \sum_{k=1}^{\infty} |(x, \varphi_k)| \|(B - B_N)\varphi_k\| \\
&\leq \left(\sum_{k=1}^{\infty} |(x, \varphi_k)|^2 \right)^{\frac{1}{2}} \left(\sum_{k=1}^{\infty} \|(B - B_N)\varphi_k\|^2 \right)^{\frac{1}{2}} \\
&= \|x\| \left(\sum_{k=1}^{\infty} \|(B - B_N)\varphi_k\|^2 \right)^{\frac{1}{2}} .
\end{aligned}
\tag{2.6}
$$

By (2.2), (2.4) and (2.5),

$$
(B - B_N)\varphi_k = \begin{cases} \sum_{j=N}^{\infty} b_{jk}\varphi_j, & 1 \leq k \leq N, \\ \sum_{j=1}^{\infty} b_{jk}\varphi_j, & k > N. \end{cases}
\tag{2.7}
$$

So, by (2.7) and Parseval's identity,

$$
\|(B - B_N)\varphi_k\|^2 = \begin{cases} \sum_{j=N}^{\infty} |b_{jk}|^2, & 1 \leq k \leq N, \\ \sum_{j=1}^{\infty} |b_{jk}|^2, & k > N. \end{cases}
\tag{2.8}
$$

Therefore, by (2.3), (2.8) and Fubini's theorem,

$$
\begin{aligned}
\sum_{k=1}^{\infty} \|(B - B_N)\varphi_k\|^2 &= \sum_{k=1}^{N} \sum_{j=N}^{\infty} |b_{jk}|^2 + \sum_{k=N}^{\infty} \sum_{j=1}^{\infty} |b_{jk}|^2 \\
&\leq \sum_{j=N}^{\infty} \sum_{k=1}^{\infty} |b_{jk}|^2 + \sum_{k=N}^{\infty} \sum_{j=1}^{\infty} |b_{jk}|^2 \to 0
\end{aligned}
\tag{2.9}
$$

as $N \to \infty$. Thus, by (2.6) and (2.9), $B_N \to B$ in the norm of $B(X)$. Hence B is compact. Since $A = B^2$, it follows that $A : X \to X$ is compact. $\qquad\square$

We have the following criterion for a positive operator to be in the trace class S_1.

Proposition 2.4 *Let $A : X \to X$ be a positive operator such that*

$$
\sum_{k=1}^{\infty} (A\varphi_k, \varphi_k) < \infty
$$

for all orthonormal bases $\{\varphi_k : k = 1, 2, \ldots\}$ for X. Then $A : X \to X$ is in the trace class S_1.

Proof. By Proposition 2.3, $A : X \to X$ is compact. Let $\{\psi_k : k = 1, 2, \ldots\}$ be an orthonormal basis for X consisting of eigenvectors of $A : X \to X$. Let $s_k(A)$ be the eigenvalue of $A : X \to X$ corresponding to the eigenvector ψ_k, $k = 1, 2, \ldots$. Then

$$\sum_{k=1}^{\infty} s_k(A) = \sum_{k=1}^{\infty} (A\psi_k, \psi_k) < \infty,$$

and the proof is complete. $\square$

We have the following criterion for a compact operator $A : X \to X$ to be in the trace class S_1. It is worth noting that positivity is no longer required.

Proposition 2.5 *Let $A : X \to X$ be a compact operator such that*

$$\sum_{k=1}^{\infty} |(A\varphi_k, \psi_k)| < \infty$$

for all orthonormal sets $\{\varphi_k : k = 1, 2, \ldots\}$ and $\{\psi_k : k = 1, 2, \ldots\}$ in X. Then $A : X \to X$ is in S_1.

Proof. Using the canonical form for compact operators given by Theorem 2.2, we get

$$A = \sum_{k=1}^{\infty} s_k(A)(\cdot, u_k)v_k,$$

where $\{u_k : k = 1, 2, \ldots\}$ is an orthonormal basis for $N(A)^{\perp}$ consisting of eigenvectors of $|A| : X \to X$, $v_k = V u_k$, $k = 1, 2, \ldots$, and the series converges to A strongly. Thus,

$$\sum_{j=1}^{\infty} |(Au_j, v_j)| = \sum_{j=1}^{\infty} s_j(A) < \infty,$$

and the proof is complete. $\square$

The following proposition is some sort of a converse of Proposition 2.4.

Proposition 2.6 *Let $A : X \to X$ be a bounded linear operator in the trace class S_1. Then $\sum_{k=1}^{\infty}(A\varphi_k, \varphi_k)$ is absolutely convergent for all orthonormal bases $\{\varphi_k : k = 1, 2, \ldots\}$ for X. Moreover, the sum of the series is independent of the choice of the orthonormal basis.*

Proof. Let $\{\varphi_k : k = 1, 2, \ldots\}$ be an orthonormal basis for X and let $\{\psi_k : k = 1, 2, \ldots\}$ be an orthonormal basis for X consisting of eigenvectors of $|A| : X \to X$. Then, by Parseval's identity,

$$\sum_{j=1}^{\infty}(A\varphi_j, \varphi_j) = \sum_{j=1}^{\infty}(\varphi_j, A^*\varphi_j)$$

$$= \sum_{j=1}^{\infty} \sum_{k=1}^{\infty} (\varphi_j, \psi_k)(\psi_k, A^* \varphi_j)$$

$$= \sum_{k=1}^{\infty} \sum_{j=1}^{\infty} (\varphi_j, \psi_k)(A\psi_k, \varphi_j)$$

$$= \sum_{k=1}^{\infty} (A\psi_k, \psi_k)$$

provided that the interchange of the order of summation is justified. But, by Theorem 2.1, we get $A = V|A|$, where $V : X \to X$ is a partial isometry. If we let $s_k(A)$ be the eigenvalue of $|A| : X \to X$ corresponding to $\psi_k : k = 1, 2, \ldots$, then, by Schwarz' inequality, Parseval's identity, $\|\varphi_k\| = \|\psi_k\| = 1$, $k = 1, 2, \ldots$, and $\|V\|_* = 1$, we get

$$\sum_{k=1}^{\infty} \sum_{j=1}^{\infty} |(\varphi_j, \psi_k)(A\psi_k, \varphi_j)|$$

$$\leq \sum_{k=1}^{\infty} \left\{ \sum_{j=1}^{\infty} |(\varphi_j, \psi_k)|^2 \right\}^{\frac{1}{2}} \left\{ \sum_{j=1}^{\infty} |(A\psi_k, \varphi_j)|^2 \right\}^{\frac{1}{2}}$$

$$= \sum_{k=1}^{\infty} \|A\psi_k\| = \sum_{k=1}^{\infty} \|V|A|\psi_k\|$$

$$= \sum_{k=1}^{\infty} s_k(A)\|V\psi_k\| \leq \|A\|_{S_1} < \infty,$$

and this completes the proof. $\qquad\qquad\qquad\qquad\qquad\qquad\qquad\square$

In view of Proposition 2.6, we can define the trace $\mathrm{tr}(A)$ of any bounded linear operator $A : X \to X$ in the trace class S_1 by

$$\mathrm{tr}(A) = \sum_{k=1}^{\infty} (A\varphi_k, \varphi_k),$$

where $\{\varphi_k : k = 1, 2, \ldots\}$ is any orthonormal basis for X.

Proposition 2.7 *Let $A : X \to X$ be a positive operator in the trace class S_1. Then*

$$\|A\|_{S_1} = \mathrm{tr}(A).$$

Proof. Using the definition of $\mathrm{tr}(A)$,

$$\mathrm{tr}(A) = \sum_{k=1}^{\infty} (A\psi_k, \psi_k), \qquad\qquad\qquad (2.10)$$

where $\{\psi_k : k = 1, 2, \ldots\}$ is an orthonormal basis for X consisting of eigenvectors of $A : X \to X$. If $s_k(A)$ is the eigenvalue of $A : X \to X$ corresponding to the eigenvector ψ_k, $k = 1, 2, \ldots$, then, by (2.10),

$$\operatorname{tr}(A) = \sum_{k=1}^{\infty} s_k(A) = \|A\|_{S_1},$$

and the proof is complete. $\square$

The following theorem gives a criterion for a bounded linear operator $A : X \to X$ to be in the Hilbert-Schmidt class S_2 and a formula for the norm $\|A\|_{S_2}$ of $A : X \to X$ in S_2.

Proposition 2.8 *Let $A : X \to X$ be a bounded linear operator such that*

$$\sum_{k=1}^{\infty} \|A\varphi_k\|^2 < \infty$$

for all orthonormal bases $\{\varphi_k : k = 1, 2, \ldots\}$ for X. Then $A : X \to X$ is in the Hilbert-Schmidt class S_2 and

$$\|A\|_{S_2}^2 = \sum_{k=1}^{\infty} \|A\varphi_k\|^2,$$

where $\{\varphi_k : k = 1, 2, \ldots\}$ is any orthonormal basis for X.

We need the following lemma in the proof of Proposition 2.8.

Lemma 2.9 *Let $\{\varphi_k : k = 1, 2, \ldots\}$ be an orthonormal basis for X consisting of eigenvectors of a bounded linear operator $A : X \to X$. For $k = 1, 2, \ldots$, let λ_k be the eigenvalue of $A : X \to X$ corresponding to φ_k. Then the spectrum of $A : X \to X$ is the closure in $\mathbb{C}$ of the set $\{\lambda_k : k = 1, 2, \ldots\}$.*

Proof. Let λ be any complex number which is not in the closure of the set $\{\lambda_k : k = 1, 2, \ldots\}$. Then there exists a positive number δ such that

$$|\lambda_k - \lambda| \geq \delta, \quad k = 1, 2, \ldots.$$

Let $y \in X$. Then let $x \in X$ be defined by

$$x = \sum_{k=1}^{\infty} \frac{(y, \varphi_k)}{\lambda_k - \lambda} \varphi_k, \tag{2.11}$$

where the convergence of the series is understood to be in X. By (2.11) and Parseval's identity,

$$(A - \lambda I)x = \sum_{k=1}^{\infty} \frac{(y, \varphi_k)}{\lambda_k - \lambda}(A - \lambda I)\varphi_k = \sum_{k=1}^{\infty} (y, \varphi_k)\varphi_k = y,$$

where I is the identity operator on X. Therefore the linear operator $A - \lambda I : X \to X$ is onto. Moreover, for any x in X, we get, by Parseval's identity,

$$\|(A - \lambda I)x\| = \left\{ \sum_{k=1}^{\infty} |\lambda_k - \lambda|^2 |(x, \varphi_k)|^2 \right\}^{\frac{1}{2}}$$

$$\geq \delta \left\{ \sum_{k=1}^{\infty} |(x, \varphi_k)|^2 \right\}^{\frac{1}{2}} = \delta \|x\|.$$

Therefore λ belongs to the resolvent set of $A : X \to X$, and the proof is complete.
□

Proof of Proposition 2.8. Let $B = A^*A$. Then $B : X \to X$ is a positive operator. Since

$$\sum_{k=1}^{\infty} (B\varphi_k, \varphi_k) = \sum_{k=1}^{\infty} (A^*\varphi_k, \varphi_k) = \sum_{k=1}^{\infty} \|A\varphi_k\|^2 < \infty$$

for all orthonormal bases $\{\varphi_k : k = 1, 2, \ldots\}$, it follows from Proposition 2.4 that $B : X \to X$ is in the trace class S_1. Since $|A| = \sqrt{B}$, it follows that $|A| : X \to X$ is compact. By Theorem 2.1, we can write $A = V|A|$, where $V : X \to X$ is a partial isometry. Therefore $A : X \to X$ is compact. Let $\{\psi_k : k = 1, 2, \ldots\}$ be an orthonormal basis for X consisting of eigenvectors of $|A| : X \to X$ and for $k = 1, 2, \ldots$, let $s_k(A)$ be the eigenvalue of $|A| : X \to X$ corresponding to ψ_k. Then

$$B\psi_k = |A|^2 \psi_k = (s_k(A))^2 \psi_k, \quad k = 1, 2, \ldots.$$

So, by Lemma 2.9, the spectrum of $B : X \to X$ is the closure in $\mathbb{C}$ of the set $\{(s_k(A))^2 : k = 1, 2, \ldots\}$. Thus,

$$\sum_{k=1}^{\infty} (s_k(A))^2 = \sum_{k=1}^{\infty} s_k(B) < \infty,$$

and hence $A : X \to X$ is in the Hilbert-Schmidt class S_2. By Propositions 2.6 and 2.7, we get

$$\|A\|_{S_2} = \sum_{k=1}^{\infty} (s_k(A))^2 = \sum_{k=1}^{\infty} s_k(B)$$

$$= \|B\|_{S_1} = \operatorname{tr}(B) = \sum_{k=1}^{\infty} (B\varphi_k, \varphi_k)$$

$$= \sum_{k=1}^{\infty} (A^*A\varphi_k, \varphi_k) = \sum_{k=1}^{\infty} \|A\varphi_k\|^2,$$

where $\{\varphi_k : k = 1, 2, \ldots\}$ is any orthonormal basis for X. This completes the proof.
□

In order to obtain some information on the Schatten-von Neumann class S_p, $1 \leq p \leq \infty$, we need interpolation theory, which we now recall. Good references for interpolation theory include the book [5] by Bergh and Löfström, the book [77] by Schechter and the book [108] by Zhu.

Let B_0 and B_1 be complex Banach spaces in which the norms are denoted by $\| \; \|_{B_0}$ and $\| \; \|_{B_1}$ respectively. We say that B_0 and B_1 are compatible if there is a complex vector space V such that $B_0 \subseteq V$ and $B_1 \subseteq V$. If this is the case, then the subspaces $B_0 \cap B_1$ and $B_0 + B_1$ of V are complex Banach spaces when equipped with the norms $\| \; \|_{B_0 \cap B_1}$ and $\| \; \|_{B_0+B_1}$ given by

$$\|v\|_{B_0 \cap B_1} = \max_{k=0,1} \|v\|_{B_k}$$

for all v in $B_0 \cap B_1$, and

$$\|v\|_{B_0+B_1} = \inf\{\|b_0\|_{B_0} + \|b_1\|_{B_1} : v = b_0 + b_1, b_0 \in B_0, b_1 \in B_1\}$$

for all v in $B_0 + B_1$, respectively.

Let B_0 and B_1 be compatible Banach spaces. A complex Banach space B is called an intermediate space between B_0 and B_1 if

$$B_0 \cap B_1 \subseteq B \subseteq B_0 + B_1,$$

where the inclusions are continuous. An intermediate space B between B_0 and B_1 is said to be an interpolation space between B_0 and B_1 if any bounded linear operator on $B_0 + B_1$, which is bounded from B_k into B_k, $k = 0, 1$, is also bounded from B into B.

Let $S = \{z \in \mathbb{C} : 0 < \mathrm{Re}\, z < 1\}$ and let B be any complex Banach space. A function $f : S \to B$ is said to be analytic on S if for any bounded linear functional b' on B, the complex-valued function $b' \circ f : S \to \mathbb{C}$ is analytic on S.

Let B_0 and B_1 be compatible Banach spaces. Then we define $\mathcal{F}(B_0, B_1)$ to be the set of all bounded and continuous functions f from the closure $\overline{S}$ of S into $B_0 + B_1$ such that f is analytic on S and the mappings

$$\mathbb{R} \ni y \mapsto f(k + iy) \in B_k, \quad k = 0, 1,$$

are continuous from $\mathbb{R}$ into B_k, $k = 0, 1$. Then it can be shown that $\mathcal{F}(B_0, B_1)$ is a complex Banach space with respect to the norm $\| \; \|_{\mathcal{F}}$ given by

$$\|f\|_{\mathcal{F}} = \max_{k=0,1} \sup_{y \in \mathbb{R}} \|f(k + iy)\|_{B_k}, \quad f \in \mathcal{F}(B_0, B_1).$$

For any number θ in $[0, 1]$, we let B_θ be the subspace of $B_0 + B_1$ consisting of all elements b in $B_0 + B_1$ such that $b = f(\theta)$ for some f in $\mathcal{F}(B_0, B_1)$. Then we can show that B_θ is a complex Banach space with respect to the norm $\|\ \|_\theta$ given by

$$\|b\|_{B_\theta} = \inf_{b=f(\theta)} \|f\|_{\mathcal{F}}, \quad b \in B_\theta,$$

and B_θ is an interpolation space between B_0 and B_1. We denote B_θ by $[B_0, B_1]_\theta$.

We have the following result on the boundedness of linear operators from an interpolation space between a pair of compatible Banach spaces into the corresponding interpolation space between another pair of compatible Banach spaces.

Theorem 2.10 *Let B_0, B_1 and $\tilde{B}_0$, $\tilde{B}_1$ be two pairs of compatible Banach spaces. Let A be a bounded linear operator from $B_0 + B_1$ into $\tilde{B}_0 + \tilde{B}_1$ such that A is a bounded linear operator from B_k into $\tilde{B}_k$ with norm $\leq M_k$, $k = 0, 1$. Then for any number θ in $(0, 1)$, A is a bounded linear operator from $[B_0, B_1]_\theta$ into $[\tilde{B}_0, \tilde{B}_1]_\theta$ with norm $\leq M_0^{1-\theta} M_1^{\theta}$.*

The Lebesgue space $L^p(M, \mu)$, where (M, μ) is a measure space, and the Schatten-von Neumann class S_p, $1 \leq p \leq \infty$, are standard examples of interpolation spaces. These facts are made precise by the following theorem.

Theorem 2.11 *For $1 \leq p \leq \infty$,*

$$[L^1(M, \mu), L^\infty(M, \mu)]_{\frac{1}{p'}} = L^p(M, \mu)$$

and

$$[S_1, S_\infty]_{\frac{1}{p'}} = S_p,$$

where (M, μ) is a measure space and p' is the conjugate index of p.

3 Topological Groups

This chapter contains the basic information on topological groups. Analysis on topological groups requires a study of Haar measures and modular functions, which we give in the next chapter. Basic references include Folland [27] and Pontryagin [68]. The book [45] is a good reference for the basic group theory used in this book. As for general topology, the books [54] and [64] by Kelley and Munkres respectively are standard references.

Let G be a group on which the binary operation is denoted by $\cdot$. Suppose that G is also a topological space such that the mappings $G \times G \ni (g, h) \mapsto g \cdot h \in G$ and $G \ni g \mapsto g^{-1} \in G$ are continuous, where g^{-1} is the inverse of g. Then we call G a topological group.

Some remarks on the definition of a topological group are in order.

Remark 3.1 The continuity of the mapping $G \times G \ni (g, h) \mapsto g \cdot h \in G$ means that for all g and h in G and any neighborhood W of $g \cdot h$, we can find a neighborhood U of g and a neighborhood V of h such that $u \cdot v \in W$ for all u in U and all v in V.

Remark 3.2 The continuity of the mapping $G \ni g \mapsto g^{-1} \in G$ means that for all g in G and any neighborhood V of g^{-1}, there exists a neighborhood U of g such that
$$u \in U \Rightarrow u^{-1} \in V.$$

Remark 3.3 For all g and h in G, $g \cdot h$ is also denoted by gh.

Theorem 3.4 *Let G be a topological group. Then G is a T_0-space $\Leftrightarrow$ G is a T_1-space $\Leftrightarrow$ G is a T_2-space $\Leftrightarrow$*

$$\bigcap_{V \in \tau, e \in V} V = \{e\}, \tag{3.1}$$

where τ is the topology in G and e is the identity element in G.

Let us recall that G is a T_0-space if for any two distinct elements in G, there exists an open set that contains exactly one of them. G is a T_1-space if for any two distinct elements g and h in G, there exist two open sets U and V such that $g \in U$, $h \in V$, $g \notin V$ and $h \notin U$. G is a T_2-space means that for any two distinct elements g and h in G, we can find open sets U and V for which $g \in U$, $h \in V$ and $U \cap V = \phi$. A T_2-space is also known as a Hausdorff space.

Before the proof of Theorem 3.4, we introduce some notation. Let $U \subseteq G$. Then we denote the set $\{u^{-1} : u \in U\}$ by U^{-1}. For all elements g and h in G, we let gU and Uh be the sets defined by

$$gU = \{gu : u \in U\} \qquad \text{and} \qquad Uh = \{uh : u \in U\}.$$

If $U \subseteq G$ and $V \subseteq G$, then we define $U \cdot V$ by

$$U \cdot V = \{uv : u \in U, v \in V\}.$$

Proof of Theorem 3.4. Suppose that G is a T_0-space and let g and h be two distinct elements in G. Then we can find an open set U that contains either g or h, but not both. We assume that $g \in U$ and $h \notin U$. Let

$$V = hU^{-1}g = \{hu^{-1}g : u \in U\}.$$

Then it is easy to see that V is an open set and $h \in V$. Now, $g \notin V$. Indeed, if $g \in V$, then there exists an element u in U such that $g = hu^{-1}g$. Thus, $h = u \in U$ and this is a contradiction. Hence G is a T_1-space. Next, suppose that G is a T_1-space and let g and h be two distinct elements in G. Let $H = G - \{h^{-1}g\}$, i.e., H is the complement of the set $\{h^{-1}g\}$ in G. Then $e \in H$. Let $u \in H$. Then $u \neq h^{-1}g$. Since G is a T_1-space, it follows that there exist open sets O_1 and O_2 such that $u \in O_1$, $h^{-1}g \in O_2$, $u \notin O_2$ and $h^{-1}g \notin O_1$. Thus, O_1 is an open neighborhood of u, which is contained in H. Therefore every point in H is an interior point. Hence H is an open set. Using the continuity of the multiplication and the fact that $ee = e$, there exist open neighborhoods U and V of e such that

$$u \in U, v \in V \Rightarrow uv \in H.$$

Let

$$W = U \cap U^{-1} \cap V \cap V^{-1}.$$

Then W is an open set and $e \in W$. Furthermore, $W = W^{-1}$ and

$$W \cdot W^{-1} \subseteq H.$$

If there exist elements s and t in W such that $gs = ht$, then $g = hts^{-1} \in hH = G - \{g\}$, and this is impossible. Therefore $gW \cap hW = \phi$. In other words, G is a T_2-space. Now, suppose that G is a T_2-space. Then for all g in G such that $g \neq e$, there exist open sets U and V for which $g \in U$, $e \in V$ and $U \cap V = \phi$. Hence (3.1) is proved. Finally, suppose that (3.1) is valid and let g and h be two distinct elements in G. Then $g^{-1}h \neq e$. So, there exists an open set U such that $e \in U$ and $g^{-1}h \notin U$. Therefore $g \in gU$ and $h \notin gU$. This proves that G is a T_0-space, and the proof of Theorem 3.4 is complete. $\qquad\square$

We assume that all topological groups in this book are T_2-topological groups, which we simply call Hausdorff groups. In view of Theorem 3.4, this is not a severe restriction.

Let G be a Hausdorff group. Let f be a complex-valued function on G. Suppose that for all positive numbers ε, there exists a neighborhood U of e such that

$$g^{-1}h \in U \Rightarrow |f(g) - f(h)| < \varepsilon.$$

Then we say that f is left uniformly continuous on G. If we replace $g^{-1}h \in U$ by $hg^{-1} \in U$ in the definition, then we have the notion of right uniform continuity on G.

Remark 3.5 Let U be any neighborhood of the identity element e in G. Then we can always find a neighborhood V of e such that $V \subseteq U$ and V is symmetric, *i.e.*, $V = V^{-1}$. Indeed, we can pick V to be the neighborhood $U \cap U^{-1}$. Thus,

$$g^{-1}h \in V \Leftrightarrow h^{-1}g \in V \quad \text{and} \quad hg^{-1} \in V \Leftrightarrow gh^{-1} \in V.$$

Therefore the roles of g and h in the definitions of left and right uniform continuity are symmetric.

Let f be a continuous and complex-valued function on G. Then the support $\text{supp}(f)$ of the function f is defined to be the closure of the set $\{g \in G : f(g) \neq 0\}$. We let $C_0(G)$ be the set of all continuous and complex-valued functions f on G such that $\text{supp}(f)$ is compact.

Proposition 3.6 *Let G be a Hausdorff group. Let $f \in C_0(G)$. Then f is left and right uniformly continuous on G.*

Proof. Let $K = \text{supp}(f)$. Then for all positive numbers ε and for all g in G, there exists an open set U_g such that $g \in U_g$ and

$$h \in U_g \Rightarrow |f(g) - f(h)| < \varepsilon.$$

Let $W_g = g^{-1}U_g$. Then W_g is an open set and $e \in W_g$. By Remark 3.5, we can find a symmetric and open set V_g of e such that $V_g \subseteq W_g$ and $V_g \cdot V_g \subseteq W_g$. Now, $\{gV_g : g \in K\}$ is an open cover of K. Since K is compact, we can find elements $g_1, g_2, \ldots, g_N$ in K such that

$$K \subset \bigcup_{j=1}^{N} (g_j V_{g_j}).$$

Let

$$V = \bigcap_{j=1}^{N} V_{g_j}.$$

Then V is a symmetric and open neighborhood of e. Now, let g and h be elements in G. If $g \in K$ and $g^{-1}h \in V$, then there exists an integer j, $j = 1, 2, \ldots, N$, such that $g \in g_j V_{g_j}$. So, on one hand, $g \in g_j W_{g_j} = U_{g_j}$, and hence

$$|f(g) - f(g_j)| < \varepsilon. \tag{3.2}$$

On the other hand, we get $h \in gV \subseteq (g_j V_{g_j}) \cdot V \subseteq g_j W_{g_j} = U_{g_j}$, and hence

$$|f(h) - f(g_j)| < \varepsilon. \tag{3.3}$$

Thus, by (3.2) and (3.3), we get

$$|f(g) - f(h)| < 2\varepsilon.$$

If $h \in K$ and $g^{-1}h \in V$, then, by Remark 3.5, $h^{-1}g \in V$, and by what we have just proved, we also get $|f(g) - f(h)| < 2\varepsilon$. If $g^{-1}h \in V$, $g \notin K$ and $h \notin K$, then it is obvious that $f(g) = f(h) = 0$. Hence we have proved that f is left uniformly continuous on G. The proof for right uniform continuity is similar and hence omitted.   □

A complex-valued function on G that is left and right uniformly continuous on G is said to be uniformly continuous on G.

4 Haar Measures and Modular Functions

We assume a basic knowledge of measure theory on locally compact and Hausdorff topological spaces, which can be found in [73] by Royden and [74] by Rudin, among others.

Let μ be a nonzero Radon measure on a locally compact and Hausdorff group G. This means that μ is a nonzero Borel measure on G such that $\mu(K) < \infty$ for all compact subsets K of G,

$$\mu(B) = \inf_{U \in \tau, B \subseteq U} \mu(U) \tag{4.1}$$

for all Borel subsets B of G and

$$\mu(U) = \sup_{K \subseteq U} \mu(U) \tag{4.2}$$

for all U in the topology τ, where the supremum in (4.2) is taken over all compact sets K contained in U. If

$$\mu(gB) = \mu(B) \tag{4.3}$$

for all g in G and all Borel subsets B of G, then we call μ a left Haar measure on G. If

$$\mu(Bg) = \mu(B) \tag{4.4}$$

for all g in G and all Borel subsets B of G, then we call μ a right Haar measure on G. A Radon measure that is both a left and right Haar measure on G is said to be a Haar measure on G.

Remark 4.1 Conditions (4.1) and (4.2) are, respectively, known as the outer regularity and the inner regularity of the measure μ. Conditions (4.3) and (4.4) give, respectively, the left invariance and the right invariance of the measure μ.

The following theorem is a fundamental result in measure theory on locally compact and Hausdorff groups. The proof is not necessary for an understanding of the contents in this book. We are content with the statement of the theorem and the proof is omitted. For a proof, see Section 2.2 of the book [27] by Folland.

Theorem 4.2 *There exists a left Haar measure on a locally compact and Hausdorff group.*

Let μ be a Radon measure on a locally compact and Hausdorff group G. If for all Borel subsets B of G, we define $\tilde{\mu}(B)$ by

$$\tilde{\mu}(B) = \mu(B^{-1}), \tag{4.5}$$

then it is easy to see that $\tilde{\mu}$ is a Radon measure on G and we have the following simple proposition.

Proposition 4.3 μ *is a left Haar measure on G if and only if $\tilde{\mu}$ is a right Haar measure on G.*

Proof. Suppose that μ is a left Haar measure on G. Then for all g in G and all Borel sets B, we can use (4.5) and the left invariance of μ to get

$$\tilde{\mu}(Bg) = \mu((Bg)^{-1}) = \mu(g^{-1}B^{-1}) = \mu(B^{-1}) = \tilde{\mu}(B).$$

Therefore $\tilde{\mu}$ is a right Haar measure on G. Conversely, if $\tilde{\mu}$ is a right Haar measure on G, then for all g in G and all Borel sets B, we get, by (4.5),

$$\mu(gB) = \tilde{\mu}((gB)^{-1}) = \tilde{\mu}(B^{-1}g^{-1}) = \tilde{\mu}(B^{-1}) = \mu(B).$$

Therefore μ is a left Haar measure on G. $\square$

We can now give a corollary of Theorem 4.2 and Proposition 4.3.

Corollary 4.4 *There exists a right Haar measure on a locally compact and Hausdorff group.*

In order to investigate the "uniqueness" of the left Haar measure on a locally compact and Hausdorff group, we need Propositions 4.5 and 4.6.

Proposition 4.5 *Let μ be a left Haar measure on a locally compact and Hausdorff group G. Then $\mu(U) > 0$ for all nonempty and open subsets U of G, and*

$$\int_G f(g)d\mu(g) > 0$$

for all nonzero and nonnegative real-valued functions f in $C_0(G)$.

Proof. There exists a compact subset K of G such that $\mu(K) > 0$. Otherwise, $\mu(K) = 0$ for all compact sets K, and then, using the inner regularity of μ,

$$\mu(G) = \sup_{K \subseteq G} \mu(K) = 0,$$

where the supremum is taken over all compact subsets K of G. Thus, μ is the zero measure. Let U be any nonempty and open subset of G. Then $\{gU : g \in K\}$ is an open cover of K. Hence there exist $g_1, g_2, \ldots, g_N$ in G for which

$$K \subset \bigcup_{j=1}^{N} (g_j U). \tag{4.6}$$

Therefore, by (4.6) and the left invariance of μ, we get

$$0 < \mu(K) \le \sum_{j=1}^{N} \mu(g_j U) = N\mu(U).$$

Next, let f be a nonzero and nonnegative real-valued function in $C_0(G)$. Then $f(g_0) > 0$ for some $g_0 \in G$. By continuity, there exists an open neighborhood U of g_0 such that

$$g \in U \Rightarrow |f(g) - f(g_0)| < \frac{f(g_0)}{2} \Rightarrow f(g) > \frac{f(g_0)}{2}.$$

So,

$$\int_G f(g)d\mu(g) \ge \int_U f(g)d\mu(g) > \frac{f(g_0)}{2} \int_U d\mu(g) = \frac{f(g_0)}{2}\mu(U) > 0,$$

and the proof is complete. $\qquad\qquad\square$

Proposition 4.6 *Let μ be a left Haar measure on a locally compact and Hausdorff group G. Let $\varphi \in C_0(G)$. Then the function*

$$G \ni g \mapsto \int_G \varphi(kg)d\mu(k) \in \mathbb{C}$$

is continuous on G.

Proof. Let $g_0 \in G$ and let ε be any positive number. Then, by Proposition 3.6, φ is left uniformly continuous on G. So, there exists a symmetric and open neighborhood V of e such that

$$g^{-1}h \in V \Rightarrow |\varphi(g) - \varphi(h)| < \varepsilon. \qquad (4.7)$$

Since G is locally compact, it follows that the identity element e has a compact neighborhood N. Without loss of generality, we can assume that $V \subset N$. Then for all k in G and all g in $g_0 V$, we get, by (4.7),

$$(kg_0)^{-1}(kg) = g_0^{-1}g \in V$$

and hence

$$\left| \int_G (\varphi(kg_0) - \varphi(kg))d\mu(g) \right| \le \varepsilon \int_{K \cdot (N^{-1}g_0^{-1})} d\mu(g) = \varepsilon\mu(K \cdot (N^{-1}g_0^{-1})),$$

where $K = \operatorname{supp}(\varphi)$, and the proof is complete. $\qquad\qquad\square$

Theorem 4.7 *Let μ and ν be left Haar measures on a locally compact and Hausdorff group G. Then there exists a positive number a such that $\mu = a\nu$.*

Proof. Let φ be a nonzero and nonnegative real-valued function in $C_0(G)$. Then for all f in $C_0(G)$, let F be the function on $G \times G$ defined by

$$F(g,h) = \frac{f(g)\varphi(hg)}{\int_G \varphi(kg)d\nu(k)}, \quad g,h \in G. \tag{4.8}$$

Then, by Propositions 4.5 and 4.6, the denominator in (4.8) is a positive and continuous function on G and hence F is a continuous function on $G \times G$ with compact support. Using the left invariance of μ and Fubini's theorem,

$$\int_G \left(\int_G F(g,h)d\mu(g) \right) d\nu(h)$$

$$= \int_G \left(\int_G \frac{f(g)\varphi(hg)}{\int_G \varphi(kg)d\nu(k)} d\mu(g) \right) d\nu(h)$$

$$= \int_G \left(\int_G \frac{f(h^{-1}g)\varphi(g)}{\int_G \varphi(kh^{-1}g)d\nu(k)} d\mu(g) \right) d\nu(h)$$

$$= \int_G \left(\int_G \frac{f(h^{-1}g)\varphi(g)}{\int_G \varphi(kh^{-1}g)d\nu(k)} d\nu(h) \right) d\mu(g)$$

$$= \int_G \left(\int_G \frac{f(h^{-1})\varphi(g)}{\int_G \varphi(kh^{-1})d\nu(k)} d\nu(h) \right) d\mu(g)$$

$$= \left(\int_G \varphi(g)d\mu(g) \right) \left(\int_G \frac{f(h^{-1})}{\int_G \varphi(kh^{-1})d\nu(k)} d\nu(h) \right). \tag{4.9}$$

So, by (4.8), (4.9) and Fubini's theorem,

$$\int_G f(g)d\mu(g) = \left(\int_G \varphi(g)d\mu(g) \right) \left(\int_G \frac{f(h^{-1})}{\int_G \varphi(kh^{-1})d\nu(k)} d\nu(h) \right). \tag{4.10}$$

Let

$$\lambda(f,\varphi;\nu) = \int_G \frac{f(h^{-1})}{\int_G \varphi(kh^{-1})d\nu(k)} d\nu(h). \tag{4.11}$$

Then, by (4.10) and (4.11), we get

$$\int_G f(g)d\mu(g) = \lambda(f,\varphi;\nu) \int_G \varphi(g)d\mu(g). \tag{4.12}$$

Similarly,

$$\int_G f(g)d\nu(g) = \lambda(f,\varphi;\nu) \int_G \varphi(g)d\nu(g). \tag{4.13}$$

If we let a be the number defined by

$$a = \frac{\int_G \varphi(g)d\mu(g)}{\int_G \varphi(g)d\nu(g)}, \tag{4.14}$$

then, by (4.12), (4.13) and (4.14),

$$\int_G f(g)d\mu(g) = a\lambda(f,\varphi;\nu)\int_G \varphi(g)d\nu(g) = a\int_G f(g)d\nu(g). \qquad (4.15)$$

Obviously, $a > 0$ and is independent of f. Since f is an arbitrary function in $C_0(G)$, we conclude from (4.15) that $\mu = a\nu$. $\qquad \square$

Remark 4.8 For the validity of Fubini's theorem used in the proof of Theorem 4.7, we need the Haar measures μ and ν on the locally compact and Hausdorff group G to be σ-finite. The σ-finiteness is guaranteed if we assume that G is σ-compact, which means that G is a countable union of compact sets. Thus, we assume that all the locally compact and Hausdorff groups encountered in this book are σ-compact.

We can now develop the properties of Haar measures on locally compact and Hausdorff groups.

Proposition 4.9 Let μ be a left Haar measure on a locally compact and Hausdorff group G. Then μ is a finite measure if and only if G is compact.

Proof. Suppose that $\mu(G) < \infty$. By the inner regularity of μ, we can find a compact subset K of G such that $\mu(K) > 0$. Consider the family $\{gK : g \in G\}$ of compact subsets of G. Let $g_1 \in G$. If $(gK) \cap (g_1K) \neq \phi$ for all g in G, then we stop. If there exists an element g in G such that $(gK) \cap (g_1K) = \phi$, then we pick such an element and call it g_2. If $(gK) \cap \bigcup_{j=1}^{2}(g_jK) \neq \phi$ for all g in G, then we stop. Otherwise, we pick an element g_3 in G such that g_1K, g_2K and g_3K are pairwise disjoint. Repeating this argument, we can get elements $g_1, g_2, \ldots, g_N$ in G for which the sets $g_1K, g_2K, \ldots, g_NK$ are pairwise disjoint and

$$(gK) \cap \bigcup_{j=1}^{N}(g_jK) \neq \phi, \quad g \in G.$$

Otherwise, we can get an infinite sequence $\{g_jK\}_{j=1}^{\infty}$ of pairwise disjoint compact subsets of G, and hence, using the left invariance of μ, we get

$$\mu(G) \geq \sum_{j=1}^{\infty}\mu(g_jK) = \sum_{j=1}^{\infty}\mu(K) = \infty,$$

and consequently a contradiction. Thus,

$$G = \bigcup_{j=1}^{N}(g_jK) \cdot K^{-1},$$

and the proof is complete. $\qquad \square$

Let μ be a left Haar measure on a locally compact and Hausdorff group G. Let $g \in G$. Then the mapping $G \ni h \mapsto hg \in G$ is a homeomorphism of G onto G. If for all Borel subsets B of G, we define $\mu_g(B)$ by

$$\mu_g(B) = \mu(Bg), \tag{4.16}$$

then it can be checked easily that μ_g is also a left Haar measure on G. Therefore, by Theorem 4.7, we can find a positive number $\Delta(g)$ such that

$$\mu_g = \Delta(g)\mu. \tag{4.17}$$

If ν is another left Haar measure on G, then, by Theorem 4.7 again, there is a positive number a for which $\nu = a\mu$. Hence, by (4.16) and (4.17),

$$\nu_g = a\mu_g = a\Delta(g)\mu = \Delta(g)\nu.$$

This observation tells us that the positive function Δ on G is independent of the choice of the left Haar measure μ. It is determined by the group G and is called the modular function on G.

Let f be a complex-valued function on G. For all elements g in G, we define the right translation $R_g f$ of f by g by

$$(R_g f)(h) = f(hg^{-1}), \quad h \in G. \tag{4.18}$$

Now, let B be a Borel subset of G. Then for all g in G,

$$R_g \chi_B = \chi_{Bg}, \tag{4.19}$$

where χ_B and χ_{Bg} are the characteristic functions on B and Bg respectively. Indeed, for all h in G, we get, by (4.18),

$$(R_g \chi_B)(h) = \chi_B(hg^{-1}) = \begin{cases} 1, & hg^{-1} \in B, \\ 0, & hg^{-1} \notin B, \end{cases} = \begin{cases} 1, & h \in Bg, \\ 0, & h \notin Bg, \end{cases} = \chi_{Bg}(h).$$

Consequently, by (4.16), (4.17) and (4.19),

$$\int_G (R_g \chi_B)(h) d\mu(h) = \mu(Bg) = \Delta(g)\mu(B) = \Delta(g) \int_G \chi_B(h) d\mu(h). \tag{4.20}$$

Therefore for all functions f in $L^1(G, \mu)$, we get from (4.20)

$$\int_G (R_g f)(h) d\mu(h) = \Delta(g) \int_G f(h) d\mu(h)$$

or, by (4.18),

$$\int_G f(hg^{-1}) d\mu(h) = \Delta(g) \int_G f(h) d\mu(h). \tag{4.21}$$

Proposition 4.10 *The modular function Δ on a locally compact and Hausdorff group G is a continuous function on G such that*

$$\Delta(gh) = \Delta(g)\Delta(h), \quad g, h \in G.$$

Proposition 4.10 tells us that the modular function $\Delta : G \to \mathbb{R}^\times$ is a group homomorphism, where $\mathbb{R}^\times$ is the group of all positive numbers with respect to multiplication.

Proof of Proposition 4.10. Let μ be a left Haar measure on G. Let $f \in C_0(G)$ be such that

$$\int_G f(h)d\mu(h) = 1.$$

Then, by (4.21) and Proposition 4.6, $\Delta : G \to (0, \infty)$ is continuous. Now, for any Borel subset B of G, we get, by (4.16) and (4.17),

$$\Delta(gh)\mu(B) = \mu(Bgh) = \Delta(h)\mu(Bg) = \Delta(h)\Delta(g)\mu(B) \qquad (4.22)$$

for all g and h in G. Thus, by (4.22),

$$\Delta(gh) = \Delta(g)\Delta(h), \quad g, h \in G,$$

and the proof of Proposition 4.10 is complete. $\qquad\qquad\qquad\square$

A locally compact and Hausdorff group G is said to be unimodular if

$$\Delta(g) = 1, \quad g \in G.$$

Proposition 4.11 *Let μ be a left Haar measure on a locally compact and Hausdorff group G. Then G is unimodular if and only if $\tilde{\mu} = \mu$, where $\tilde{\mu}$ is defined by (4.5).*

We need the following lemma for the proof of Proposition 4.11.

Lemma 4.12 *Let μ be a left Haar measure and let Δ be the modular function on a locally compact and Hausdorff group G. Then for all Borel subsets B of G,*

$$\tilde{\mu}(B) = \int_B \Delta(g^{-1})d\mu(g),$$

where $\tilde{\mu}$ is defined by (4.5).

Proof. Let $f \in C_0(G)$ be such that

$$\int_G f(h)d\mu(h) = 1. \qquad (4.23)$$

Then, by (4.21) and (4.23),

$$\Delta(g^{-1}) = \int_G f(hg)d\mu(h),$$

and hence for all Borel subsets B of G, we can use Fubini's theorem, the left invariance of μ and (4.5) to get

$$\int_B \Delta(g^{-1})d\mu(g)$$

$$= \int_G \chi_B(g)\Delta(g^{-1})d\mu(g)$$

$$= \int_G \left(\int_G \chi_B(g)f(hg)d\mu(h) \right) d\mu(g)$$

$$= \int_G \left(\int_G \chi_B(g)f(hg)d\mu(g) \right) d\mu(h)$$

$$= \int_G \left(\int_G \chi_B(h^{-1}g)f(g)d\mu(g) \right) d\mu(h)$$

$$= \int_G \left(\int_G \chi_B(h^{-1}g)d\mu(h) \right) f(g)d\mu(g)$$

$$= \int_G \left(\int_G \chi_B(h^{-1})d\mu(h) \right) f(g)d\mu(g)$$

$$= \int_G \chi_{B^{-1}}(h)d\mu(h) = \mu(B^{-1}) = \tilde{\mu}(B).$$

$\square$

Proof of Proposition 4.11. Suppose that G is unimodular. Then $\Delta(g) = 1$ for all g in G. Let B be a Borel subset of G. Then, by Lemma 4.12,

$$\tilde{\mu}(B) = \int_B \Delta(g^{-1})d\mu(g) = \int_B d\mu(g) = \mu(B).$$

Thus, $\tilde{\mu} = \mu$. Conversely, suppose that $\tilde{\mu} = \mu$. If $\Delta(g_0) \neq 1$ for some g_0 in G. Then, by continuity, we can find an open neighborhood U of g_0 such that $\Delta(g^{-1}) \neq 1$ for all g in U. To be specific, we assume that $\Delta(g^{-1}) > 1$ for all g in U. Therefore, by Lemma 4.12,

$$\mu(U) = \tilde{\mu}(U) = \int_U \Delta(g^{-1})d\mu(g) > \int_U d\mu(U) = \mu(U),$$

and this is a contradiction. Therefore $\Delta(g) = 1$ for all g in G, *i.e.*, G is unimodular.

$\square$

Proposition 4.13 *Every abelian, locally compact and Hausdorff group is unimodular.*

Proof. Let $g \in G$ and let B be any Borel subset of G. Then, using (4.16), the assumption that G is abelian and the left invariance of μ, we get

$$\mu_g(B) = \mu(Bg) = \mu(gB) = \mu(B). \tag{4.24}$$

On the other hand, by (4.17),

$$\mu_g(B) = \Delta(g)\mu(B).$$ (4.25)

By (4.24) and (4.25), we get

$$\Delta(g)\mu(B) = \mu(B),$$

and hence $\Delta(g) = 1$ for all g in G. Therefore G is unimodular. □

Proposition 4.14 *Every compact and Hausdorff group is unimodular.*

Proof. Let Δ be the modular function on G. Then, using the fact that $\Delta : G \to \mathbb{R}^\times$ is a group homomorphism, we get

$$\Delta(g^n) = (\Delta(g))^n, \quad g \in G,$$ (4.26)

for all positive integers n. If there exists an element g_0 in G for which $\Delta(g_0) \neq 1$, then $\Delta(g_0) > 1$ or $\Delta(g_0^{-1}) > 1$. Let us assume that $\Delta(g_0) > 1$. Then, by (4.26), $\Delta(g_0^n) \to \infty$ as $n \to \infty$. This contradicts the fact that Δ is a continuous function on the compact group G. Therefore $\Delta(g) = 1$ for all g in G, *i.e.*, G is unimodular. □

The Weyl-Heisenberg group and the Heisenberg group, to be studied in Chapter 17, are concrete unimodular groups, which are neither abelian nor compact. The affine group, to be studied in Chapter 18, is a non-unimodular group.

We assume throughout the book that a locally compact and Hausdorff group G is always equipped with a left Haar measure, which we denote by μ.

5 Unitary Representations

This chapter is a brief account on unitary representations of locally compact and Hausdorff groups on separable and complex Hilbert spaces. Only the most basic topics are touched on in this chapter. The more advanced theory of square-integrable representations is given in the next chapter. A good reference for this chapter is Chapter 3 of the book [27] by Folland. A more comprehensive treatise is the book [55] by Kirillov.

Let G be a locally compact and Hausdorff group. Let X be a separable and complex Hilbert space in which the inner product and the norm are denoted by $(\,,)$ and $\|\,\|$ respectively. The group of all unitary operators on X with respect to the usual composition of mappings is denoted by $U(X)$. A group homomorphism $\pi : G \to U(X)$ is said to be a unitary representation of G on X if it is strongly continuous, *i.e.*, $G \ni g \mapsto \pi(g)x \in X$ is a continuous mapping for all x in X. The Hilbert space X is called the representation space of $\pi : G \to U(X)$ and the dimension of X is known as the dimension or the degree of $\pi : G \to U(X)$.

Remark 5.1 We can replace the requirement that $\pi : G \to U(X)$ be strongly continuous by the weaker condition of weak continuity, *i.e.*, the condition that the function $G \ni g \mapsto (\pi(g)x, y) \in \mathbb{C}$ be continuous for all x and y in X. Indeed, let $\{g_j\}_{j \in J}$ be a net in G such that $g_j \to g$ for some g in G, then

$$(\pi(g_j)x, y) \to (\pi(g)x, y) \tag{5.1}$$

for all x and y in X. Since $\pi(g)$ and $\pi(g_j)$, $j \in J$, are unitary operators on X, it follows from (5.1) that

$$
\begin{aligned}
\|\pi(g_j)x - \pi(g)x\|^2 &= \|\pi(g_j)x\|^2 + \|\pi(g)x\|^2 - 2\mathrm{Re}(\pi(g_j)x, \pi(g)x) \\
&= 2\|x\|^2 - 2\mathrm{Re}(\pi(g_j)x, \pi(g)x) \\
&\to 2\|x\|^2 - 2\|\pi(g)x\|^2 = 2\|x\|^2 - 2\|x\|^2 = 0.
\end{aligned}
$$

Thus, $\pi : G \to U(X)$ is strongly continuous.

A closed subspace M of X is said to be invariant with respect to the unitary representation $\pi : G \to U(X)$ of G on X if $\pi(g)M \subseteq M$ for all g in G. $\{0\}$ and X are the trivial invariant subspaces. It is important to emphasize that all invariant subspaces are closed by definition. A unitary representation $\pi : G \to U(X)$ of G on X is said to be irreducible if it has only the trivial invariant subspaces.

A fundamental result in representation theory is the following theorem, which is usually referred to as Schur's lemma.

Theorem 5.2 *A unitary representation $\pi : G \to U(X)$ of a locally compact and Hausdorff group G on a separable and complex Hilbert space X is irreducible if and only if the only bounded linear operators on X that commute with $\pi(g)$ for all g in G are scalar multiples of the identity operator on X.*

We need the following lemma to prove Theorem 5.2.

Lemma 5.3 *Let M be an invariant subspace of X with respect to the unitary representation $\pi : G \to U(X)$ of G on X. Then the same is true for the orthogonal complement $M^\perp$ of M in X.*

Proof. Let $x \in M^\perp$. Then for all g in G and all y in M, we can use the fact that $\pi : G \to U(X)$ is a unitary representation to obtain

$$(\pi(g)x, y) = (x, (\pi(g))^*y) = (x, (\pi(g))^{-1}y) = (x, \pi(g^{-1})y) = 0,$$

where $(\pi(g))^*$ is the adjoint of $\pi(g)$. Therefore $\pi(g)x \in M^\perp$ for all g in G.   □

Another ingredient in the proof of Schur's lemma is the spectral theorem for self-adjoint operators on separable and complex Hilbert spaces. References for the spectral theorem abound in the literature. A good one is the book [72] by Reed and Simon.

Proof of Theorem 5.2. Suppose that $\pi : G \to U(X)$ is not irreducible. Let M be an invariant subspace of X with respect to $\pi : G \to U(X)$ such that $M \neq \{0\}$ and $M \neq X$. Let P be the orthogonal projection of X onto M. Then P is a bounded linear operator on X. Moreover, for all g in G, we get

$$\pi(g)Px = \pi(g)x = P\pi(g)x, \quad x \in M,$$

and

$$\pi(g)Px = 0 = P\pi(g)x, \quad x \in M^\perp.$$

Thus, P is a bounded linear operator on X that commutes with $\pi(g)$ for all g in G, and P is not a scalar multiple of the identity operator on X. Conversely, suppose that A is a bounded linear operator on X such that A is not a scalar multiple of the identity operator on X and A commutes with $\pi(g)$ for all g in G. Then the bounded linear operators S and T on X, defined by

$$S = \frac{1}{2}(A + A^*) \qquad \text{and} \qquad T = -\frac{1}{2}i(A - A^*),$$

where A^* is the adjoint of A, are self-adjoint. Since $\pi : G \to U(X)$ is a unitary representation of G on X for all g in G and A commutes with $\pi(g)$ for all g in G, it follows that

$$
\begin{aligned}
(A^*\pi(g)x, y) &= (\pi(g)x, Ay) = (x, (\pi(g))^* Ay) \\
&= (x, (\pi(g))^{-1} Ay) = (x, \pi(g^{-1}) Ay) \\
&= (x, A\pi(g^{-1})y) = (A^*x, \pi(g^{-1})y) = (\pi(g)A^*x, y)
\end{aligned}
$$

for all x and y in X. Thus, A^* commutes with $\pi(g)$ for all g in G. Therefore both S and T commute with $\pi(g)$ for all g in G. At least one of them is not a scalar multiple of the identity operator on X. To be specific, let us suppose that S is not a scalar multiple of the identity operator on X. Let $\{E(\lambda) : \lambda \in \mathbb{R}\}$ be the spectral family of the self-adjoint operator S. Then, by the spectral theorem, the projection $E(\lambda)$ commutes with $\pi(g)$ for all λ in $\mathbb{R}$ and all g in G. Let P be one such nonzero projection that is not the identity operator on X and let M be its range. Then M is a nontrivial closed subspace of X. Furthermore, for all g in G and all x in M, we get

$$\pi(g)x = \pi(g)Px = P\pi(g)x \in M.$$

Thus, $\pi : G \to U(X)$ is not irreducible. This completes the proof. $\square$

We can now give two consequences of Schur's lemma.

Theorem 5.4 *Let G be an abelian, locally compact and Hausdorff group. Then every irreducible and unitary representation of G on a separable and complex Hilbert space is one-dimensional.*

Proof. Let $\pi : G \to U(X)$ be an irreducible and unitary representation of G on a separable and complex Hilbert space X and let $g \in G$. Then, using the fact that G is abelian, $\pi(g)$ commutes with $\pi(h)$ for all h in G. By Schur's lemma, there exists a complex number c_g such that $\pi(g) = c_g I$, where I is the identity operator on X. Now, suppose that $\dim(X) > 1$. Let M be a closed subspace of X such that $M \neq \{0\}$ and $M \neq X$. Then for all g in G and all x in M,

$$\pi(g)x = c_g x \in M.$$

Therefore M is an invariant subspace of X and this contradicts the irreducibility of $\pi : G \to U(X)$. $\square$

Theorem 5.5 *Let G be a compact and Hausdorff group. Then every irreducible and unitary representation of G on a separable and complex Hilbert space is finite-dimensional.*

Proof. Let $\pi : G \to U(X)$ be an irreducible and unitary representation of G on X. Let $\varphi \in X$ be such that $\|\varphi\| = 1$. We define the linear operator $T_\varphi : X \to X$ by

$$(T_\varphi x, y) = \int_G (x, \pi(g)\varphi)(\pi(g)\varphi, y)d\mu(g), \quad x, y \in X. \tag{5.2}$$

Then $T_\varphi : X \to X$ is a bounded linear operator. Indeed, for all x and y in X, we get, by Schwarz' inequality, the compactness of G, $\|\varphi\| = 1$ and the fact that $\pi(g)$ is a unitary operator for all g in G,

$$|(T_\varphi x, y)| \leq \int_G |(x, \pi(g)\varphi)| \, |(\pi(g)\varphi, y)|d\mu(g) \leq \mu(G)\|x\| \, \|y\|.$$

Moreover, $T_\varphi : X \to X$ is a positive operator. Indeed, let $x \in X$. Then

$$(T_\varphi x, x) = \int_G |(x, \pi(g)\varphi)|^2 d\mu(g) \geq 0.$$

Now, let $\{\varphi_j : j = 1, 2, \ldots\}$ be an orthonormal basis for X. Then, by Fubini's theorem, Parseval's identity, $\|\varphi\| = 1$ and the fact that $\pi(g)$ is a unitary operator for all g in G, we get

$$
\begin{aligned}
\sum_{j=1}^{\infty} (T_\varphi \varphi_j, \varphi_j) &= \sum_{j=1}^{\infty} \int_G |(\varphi_j, \pi(g)\varphi)|^2 d\mu(g) \\
&= \int_G \sum_{j=1}^{\infty} |(\varphi_j, \pi(g)\varphi)|^2 d\mu(g) \\
&= \int_G \|\pi(g)\varphi\|^2 d\mu(g) = \mu(G) < \infty.
\end{aligned}
$$

Thus, by Proposition 2.4, $T_\varphi : X \to X$ is in S_1 and hence compact. Furthermore, the function $G \ni g \mapsto |(u, \pi(g)\varphi)|^2 \in \mathbb{R}$ is continuous and is equal to 1 when $g = e$. It follows that it is strictly positive in a neighborhood of e. Thus,

$$(T_\varphi \varphi, \varphi) = \int_G |(\varphi, \pi(g)\varphi)|^2 d\mu(g) > 0.$$

In other words, the compact operator $T_\varphi : X \to X$ is nonzero. Now, for all h in G, using the fact that $\pi : G \to U(X)$ is a unitary representation and the left invariance of the Haar measure μ, we get

$$
\begin{aligned}
(T_\varphi \pi(h)x, y) &= \int_G (\pi(h)x, \pi(g)\varphi)(\pi(g)\varphi, y) d\mu(g) \\
&= \int_G (x, \pi(h^{-1}g)\varphi)(\pi(g)\varphi, y) d\mu(g) \\
&= \int_G (x, \pi(g)\varphi)(\pi(hg)\varphi, y) d\mu(g) \\
&= \int_G (x, \pi(g)\varphi)(\pi(g)\varphi, \pi(h^{-1})y) d\mu(g) \\
&= (T_\varphi x, \pi(h^{-1})y) = (\pi(h)T_\varphi x, y)
\end{aligned}
$$

for all x and y in X. Thus, T_φ commutes with $\pi(g)$ for all g in G. So, using the irreducibility of $\pi : G \to U(X)$, we can find a nonzero constant c such that $T_\varphi = cI$, where I is the identity operator on X. Thus, the identity operator I on X is compact. So, X has to be finite-dimensional. $\qquad\square$

Remark 5.6 Theorem 5.5 is a well-known result in the representation theory of compact groups. See, for instance, Chapter 5 of the book [27] by Folland. It is remarkable that in the proof of Theorem 5.5, the function

$$X \ni x \mapsto (x, \pi(\cdot)\varphi) \in L^2(G)$$

in (5.2) is a wavelet transform, and the linear operator $T_\varphi : X \to X$ defined by (5.2) is a localization operator associated to the admissible wavelet φ and the symbol F on G given by

$$F(g) = 1, \quad g \in G.$$

See Chapters 7 and 12 in this connection. Thus, we have already come across wavelet transforms and localization operators in the context of the representation theory of compact groups.

The unitary representations of locally compact and Hausdorff groups that are of most interest to us in this book are infinite-dimensional. We assume from now on that all Hilbert spaces are infinite-dimensional, separable and complex.

We give an example of a unitary representation of a locally compact group on a Hilbert space.

Example 5.7 Let G be a locally compact and Hausdorff group endowed with a left Haar measure μ. Let $L : G \to U(L^2(G))$ be the mapping defined by

$$(L(g)f)(h) = f(g^{-1}h), \quad g, h \in G,$$

for all f in $L^2(G)$. Then we leave it as an exercise to prove that $L : G \to U(L^2(G))$ is a unitary representation of G on $L^2(G)$. We call it the left regular representation of G.

6 Square-Integrable Representations

We are now well-equipped to study the main contents in this book. We begin with a study of square-integrable representations of locally compact and Hausdorff groups on Hilbert spaces. This chapter can be considered as a continuation of the study of unitary representations begun in Chapter 5.

Let G be a locally compact and Hausdorff group. Let X be a Hilbert space. As before, we denote the inner product and the norm in X by $(\,,\,)$ and $\|\ \|$ respectively. A unitary representation $\pi : G \to U(X)$ of G on X is said to be square-integrable if there exists a nonzero element φ in X such that

$$\int_G |(\varphi, \pi(g)\varphi)|^2 d\mu(g) < \infty. \tag{6.1}$$

The condition (6.1) is known as the admissibility condition for the square-integrable representation of G on X. We call any element φ in X for which $\|\varphi\| = 1$ and the admissibility condition is valid an admissible wavelet for the square-integrable representation $\pi : G \to U(X)$, and we define the constant c_φ by

$$c_\varphi = \int_G |(\varphi, \pi(g)\varphi)|^2 d\mu(g). \tag{6.2}$$

We call c_φ the wavelet constant associated to the admissible wavelet φ.

Theorem 6.1 *Let $\pi : G \to U(X)$ be an irreducible and square-integrable representation of G on X. If φ is an admissible wavelet for $\pi : G \to U(X)$, then*

$$(x, y) = \frac{1}{c_\varphi} \int_G (x, \pi(g)\varphi)(\pi(g)\varphi, y) d\mu(g) \tag{6.3}$$

for all x and y in X.

Remark 6.2 In order to understand the formula (6.3) better, let us note that it tells us, informally, that

$$I = \frac{1}{c_\varphi} \int_G (\cdot, \pi(g)\varphi)\pi(g)\varphi d\mu(g),$$

where I is the identity operator on X. In other words, the identity operator I on X can be resolved into a sum of rank-one operators $\frac{1}{c_\varphi}(\cdot, \pi(g)\varphi)\pi(g)\varphi$, which are parametrized by elements g in G. Thus, the formula (6.3) is known as the resolution of the identity formula.

To prove Theorem 6.1, we need a lemma.

Lemma 6.3 *The subspace M of X defined by*

$$M = \left\{ x \in X : \int_G |(x, \pi(g)\varphi)|^2 d\mu(g) < \infty \right\} \tag{6.4}$$

is a closed subspace of X.

The role of Lemma 6.3 in the proof of Theorem 6.1 is to show that the subspace M is in fact the entire Hilbert space X and hence a closed linear operator defined on it is a bounded linear operator by the closed graph theorem.

Proof of Theorem 6.1. Using the fact that $\pi : G \to U(X)$ is a representation, the left invariance of μ and (6.4), we get

$$\int_G |(\pi(h)x, \pi(g)\varphi)|^2 d\mu(g) = \int_G |(x, \pi(h^{-1}g)\varphi)|^2 d\mu(g)$$
$$= \int_G |(x, \pi(g)\varphi)|^2 d\mu(g) < \infty$$

for all h in G and all x in M. Hence, by (6.4), $\pi(h)x \in M$ for all x in M and all h in G. Therefore, using the fact that $\varphi \in M$, M is a nonzero subspace of X, which is invariant with respect to the irreducible representation $\pi : G \to U(X)$. Since M is a closed subspace of X by Lemma 6.3, it follows that $M = X$. Now, we define the linear operator $A_\varphi : X \to L^2(G)$ by

$$(A_\varphi x)(g) = (x, \pi(g)\varphi), \quad x \in X, \ g \in G. \tag{6.5}$$

Then for $x \in X$ and $g, h \in G$, by (6.5) and the fact that $\pi : G \to U(X)$ is a representation, we have

$$(A_\varphi \pi(h)x)(g) = (\pi(h)x, \pi(g)\varphi) = (x, \pi(h^{-1}g)\varphi) = (A_\varphi x)(h^{-1}g),$$

and so,

$$A_\varphi \pi(h) = L(h)A_\varphi, \tag{6.6}$$

where $L : G \to U(L^2(G))$ is the left regular representation of G in Example 5.7, i.e.,

$$(L(h)f)(g) = f(h^{-1}g), \quad g, h \in G, \tag{6.7}$$

for all f in $L^2(G)$. Let $\{x_k\}_{k=1}^\infty$ be a sequence of elements in X such that $x_k \to x$ in X and $A_\varphi x_k \to f$ in $L^2(G)$ as $k \to \infty$. Then there is a subsequence of $\{A_\varphi x_k\}_{k=1}^\infty$, again denoted by $\{A_\varphi x_k\}_{k=1}^\infty$, such that

$$A_\varphi x_k \to f \tag{6.8}$$

a.e. on G as $k \to \infty$. Since

$$(x_k, \pi(g)\varphi) \to (x, \pi(g)\varphi), \quad g \in G,$$

as $k \to \infty$, it follows from (6.5) that

$$(A_\varphi x_k)(g) \to (A_\varphi x)(g), \quad g \in G. \tag{6.9}$$

Thus, by (6.8) and (6.9), $A_\varphi x = f$. Hence $A_\varphi : X \to L^2(G)$ is a closed linear operator, and, by the closed graph theorem, $A_\varphi : X \to L^2(G)$ is a bounded linear operator. Finally, for all x and y in X, we get, by (6.5)–(6.7) and the left invariance of μ,

$$
\begin{aligned}
(A_\varphi^* L(g) A_\varphi x, y) &= (L(g) A_\varphi x, A_\varphi y)_{L^2(G)} \\
&= \int_G (A_\varphi x)(g^{-1}h)\overline{(A_\varphi y)(h)}d\mu(h) \\
&= \int_G (A_\varphi x)(h)\overline{(A_\varphi y)(gh)}d\mu(h) \\
&= (A_\varphi x, L(g^{-1})A_\varphi y)_{L^2(G)} \\
&= (A_\varphi x, A_\varphi \pi(g^{-1})y)_{L^2(G)} \\
&= (\pi(g)A_\varphi^* A_\varphi x, y)
\end{aligned}
$$

for all g in G, where A_φ^* is the adjoint of A_φ, and hence

$$A_\varphi^* L(g) A_\varphi = \pi(g) A_\varphi^* A_\varphi, \quad g \in G. \tag{6.10}$$

Moreover, by (6.6),

$$A_\varphi^* L(g) A_\varphi = A_\varphi^* A_\varphi \pi(g), \quad g \in G. \tag{6.11}$$

Thus, by (6.10), (6.11) and the fact that $\pi : G \to U(X)$ is irreducible, we can use Schur's lemma to conclude that there exists a constant c such that

$$A_\varphi^* A_\varphi = cI, \tag{6.12}$$

where I is the identity operator on X. Thus, for all x and y in X, we get, by (6.12),

$$
\begin{aligned}
c(x, y) &= (A_\varphi^* A_\varphi x, y) = (A_\varphi x, A_\varphi y)_{L^2(G)} \\
&= \int_G (x, \pi(g)\varphi)\overline{(y, \pi(g)\varphi)}d\mu(g) \\
&= \int_G (x, \pi(g)\varphi)(\pi(g)\varphi, y)d\mu(g). \tag{6.13}
\end{aligned}
$$

So, by (6.2) and (6.13),

$$c = c(\varphi, \varphi) = \int_G |(\varphi, \pi(g)\varphi)|^2 d\mu(g) = c_\varphi. \tag{6.14}$$

Hence, by (6.13) and (6.14), the proof is complete provided that we can prove Lemma 6.3. ☐

To prove Lemma 6.3, we use the extended Schur's lemma used by Grossmann, Morlet and Paul in the paper [36]. We formulate it as Theorem 6.4 and we omit the proof.

Theorem 6.4 *Let G be a locally compact and Hausdorff group. Let X_1 and X_2 be Hilbert spaces in which the norms are denoted by $\| \ \|_{X_1}$ and $\| \ \|_{X_2}$ respectively. Let $\pi_1 : G \to U(X_1)$ be an irreducible and unitary representation of G on X_1 and let $\pi_2 : G \to U(X_2)$ be a unitary representation of G on X_2. Let A be a closed linear operator from X_1 into X_2 such that the domain $\mathcal{D}(A)$ of A is dense in X_1 and*

$$A\pi_1(g)x = \pi_2(g)Ax, \quad x \in \mathcal{D}(A).$$

Then A is a scalar multiple of an isometry from $\mathcal{D}(A)$ into X_2, where $\mathcal{D}(A)$ is the Hilbert space equipped with the graph norm $\| \ \|_A$ of A defined by

$$\|x\|_A^2 = \|x\|_{X_1}^2 + \|Ax\|_{X_2}^2, \quad x \in \mathcal{D}(A).$$

Proof of Lemma 6.3. As has been shown in the proof of Theorem 6.1, the linear operator A_φ from X into X with domain M is a closed linear operator. It has been shown in the proof of Theorem 6.1 that M is an invariant subspace of X with respect to $\pi : G \to U(X)$. So, for all x in the closure $\overline{M}$ of M in X, we get a sequence $\{x_k\}_{k=1}^\infty$ in M such that

$$x_k \to x \tag{6.15}$$

in X as $k \to \infty$, and hence, by (6.15),

$$\pi(g)x_k \to \pi(g)x$$

in X as $k \to \infty$ for all g in G. Thus, $\pi(g)x \in \overline{M}$ for all g in G. Therefore $\overline{M}$ is an invariant subspace of X with respect to $\pi : G \to U(X)$. Since $\varphi \in M$ and $\pi : G \to U(X)$ is irreducible, we can use Schur's lemma to conclude that $\overline{M} = X$. Therefore M is dense in X. By (6.6),

$$A_\varphi \pi(g)x = L(g)A_\varphi x, \quad x \in M,$$

where $L : G \to U(L^2(G))$ is the left regular representation of G defined by

$$(L(g)f)(h) = f(g^{-1}h), \quad h \in G,$$

for all g in G and all f in $L^2(G)$. Now, M becomes a Hilbert space, denoted by M_φ, if we equip it with the graph norm $\| \ \|_\varphi$ of A_φ given by

$$\|x\|_\varphi^2 = \|x\|^2 + \|A_\varphi x\|_{L^2(G)}^2, \quad x \in M. \tag{6.16}$$

Thus, by the extended Schur's lemma, we can conclude that A_φ is a scalar multiple of an isometry from M_φ into $L^2(G)$. So, by (6.16), there exists a positive constant λ such that

$$\|A_\varphi x\|_{L^2(G)}^2 = \lambda\|x\|_\varphi^2 = \lambda\|x\|^2 + \lambda\|A_\varphi x\|_{L^2(G)}^2, \quad x \in M. \tag{6.17}$$

Hence, by (6.17), $\lambda < 1$ and

$$\|A_\varphi x\|_{L^2(G)}^2 = \frac{\lambda}{1-\lambda}\|x\|^2, \quad x \in M. \qquad (6.18)$$

Using (6.18) and the density of M in X, we can extend $A_\varphi : M \to L^2(G)$ to a bounded linear operator from X into $L^2(G)$, which we denote by $\tilde{A}_\varphi : X \to L^2(G)$. So, if $\{x_k\}_{k=1}^\infty$ is a sequence of elements in M such that $x_k \to x$ in X as $k \to \infty$, then $A_\varphi x_k \to \tilde{A}_\varphi x$ in $L^2(G)$ as $k \to \infty$. Since A_φ is a closed linear operator from X into $L^2(G)$ with domain M, it follows that $x \in M$. Therefore M is a closed subspace of X. $\qquad \square$

Remark 6.5 Theorem 6.1 is a simplified version of Theorem 3.1 in the paper [36] by Grossmann, Morlet and Paul, where the original contributions due to Duflo and Moore [24] are acknowledged. Chapter 14 of the book [14] by Dixmier is devoted to a study of square-integrable representations. See also the paper [9] by Carey in this connection.

We can give some information on the set $AW(\pi)$ of admissible wavelets associated to an irreducible and unitary representation $\pi : G \to U(X)$ for unimodular groups G.

Theorem 6.6 *Let G be a unimodular group and let $\pi : G \to U(X)$ be an irreducible and unitary representation of G on X. Then $AW(\pi) = \phi$ or $AW(\pi) = \{x \in X : \|x\| = 1\}$.*

For a proof of Theorem 6.6, we use the theory of quadratic forms, which we recall without proofs. Details can be found in Section 6 of Chapter 8 of the book [72] by Reed and Simon.

Let M be a dense subspace of X. A mapping $q : M \times M \to \mathbb{C}$ is said to be a quadratic form on X with form domain M if $q(\cdot, y)$ is linear and $q(x, \cdot)$ is conjugate linear for all x and y in M. A quadratic form $q : M \times M \to \mathbb{C}$ on X with form domain M is said to be symmetric if

$$q(x, y) = \overline{q(y, x)}, \quad x, y \in M,$$

and is said to be positive if

$$q(x, x) \geq 0, \quad x \in M.$$

A positive quadratic form $q : M \times M \to \mathbb{C}$ on X with form domain M is said to be closed if M is complete with respect to the norm $\|\ \|_q$ given by

$$\|x\|_q^2 = \|x\|^2 + q(x, x), \quad x \in M. \qquad (6.19)$$

We need the following result, which is known as the second representation theorem given on page 331 of the book [50] by Kato.

Theorem 6.7 *Let* $q : M \times M \to \mathbb{C}$ *be a symmetric, positive and closed quadratic form on* X *with form domain* M. *Then there exists a unique positive and self-adjoint operator* A *from* X *into* X *with domain* M *such that*

$$(Ax, Ay) = q(x, y), \quad x, y \in M.$$

Proof of Theorem 6.6. Let $\mathcal{D}$ be the subspace of X defined by

$$\mathcal{D} = \left\{ x \in X : \int_G |(x, \pi(g)x)|^2 d\mu(g) < \infty \right\}. \tag{6.20}$$

Suppose that $\mathcal{D} \neq \{0\}$. Let $x \in \mathcal{D}$ and $h \in G$. Then, using the fact that $\pi : G \to U(X)$ is a representation, the unimodularity of G and (6.20),

$$\int_G |(\pi(h)x, \pi(g)\pi(h)x)|^2 d\mu(g) = \int_G |(x, \pi(h^{-1}gh)x)|^2 d\mu(g)$$

$$= \int_G |(x, \pi(g)x)|^2 d\mu(g) < \infty,$$

and hence, by (6.20), $\pi(h)x \in \mathcal{D}$. Therefore $\overline{\mathcal{D}}$ is an invariant subspace of X with respect to $\pi : G \to U(X)$. Thus, $\overline{\mathcal{D}} = X$. Let $\varphi \in \mathcal{D}$ and let $A_\varphi : X \to L^2(G)$ be the linear operator defined by (6.5). Then, as has been shown in the proof of Theorem 6.1,

$$A_\varphi \pi(h) = L(h)A_\varphi, \quad h \in G, \tag{6.21}$$

where $L : G \to U(L^2(G))$ is the left regular representation of G. Thus, for all φ and ψ in $\mathcal{D}$, we get, by (6.21),

$$A_\psi^* A_\varphi \pi(h) = A_\psi^* L(h)A_\varphi, \quad h \in G, \tag{6.22}$$

where A_ψ^* is the adjoint of A_ψ. Using the argument in the derivation of (6.10), we get

$$A_\psi^* L(h)A_\varphi = \pi(h)A_\psi^* A_\varphi, \quad h \in G. \tag{6.23}$$

So, by (6.22) and (6.23),

$$A_\psi^* A_\varphi \pi(h) = \pi(h)A_\psi^* A_\varphi, \quad h \in G,$$

and hence, by Schur's lemma, there exists a constant $c_{\varphi,\psi}$ such that

$$A_\psi^* A_\varphi = c_{\varphi,\psi}I, \tag{6.24}$$

where I is the identity operator on X. Thus, by (6.5) and (6.24),

$$c_{\varphi,\psi} = \frac{1}{\|x\|^2}(A_\psi^* A_\varphi x, x) = \frac{1}{\|x\|^2}(A_\varphi x, A_\psi x)_{L^2(G)}$$

$$= \frac{1}{\|x\|^2}\int_G (x, \pi(g)\varphi)(\pi(g)\psi, x)d\mu(g) \tag{6.25}$$

for all nonzero elements x in X. Let $q : \mathcal{D} \times \mathcal{D} \to \mathbb{C}$ be the quadratic form on X with form domain $\mathcal{D}$ defined by

$$q(\varphi, \psi) = c_{\varphi, \psi}, \quad \varphi, \psi \in \mathcal{D}, \tag{6.26}$$

where $c_{\varphi, \psi}$ is given by (6.25). Then, obviously, $q : \mathcal{D} \times \mathcal{D} \to \mathbb{C}$ is symmetric and positive. Let $\{\varphi_k\}_{k=1}^{\infty}$ be a Cauchy sequence in $\mathcal{D}$ with respect to the norm $\| \ \|_q$ given by (6.19). Then $\{\varphi_k\}_{k=1}^{\infty}$ is a Cauchy sequence in X. Hence

$$\varphi_k \to \varphi \tag{6.27}$$

for some φ in X as $k \to \infty$. Now, for any x in X, and $k = 1, 2, \ldots$, we define the function f_k^x on G by

$$f_k^x(g) = (x, \pi(g)\varphi_k), \quad g \in G. \tag{6.28}$$

If $x \neq 0$, then by (6.19) and (6.28),

$$\begin{aligned}
\|f_j^x - f_k^x\|_{L^2(G)}^2 &= \frac{1}{\|x\|^2} \int_G |(x, \pi(g)(\varphi_j - \varphi_k))|^2 d\mu(g) \\
&= q(\varphi_j - \varphi_k, \varphi_j - \varphi_k) \\
&\leq \|\varphi_j - \varphi_k\|_q^2 \to 0
\end{aligned}$$

as $j, k \to \infty$. Thus, $\{f_k^x\}_{k=1}^{\infty}$ is a Cauchy sequence in $L^2(G)$ for all x in X. So,

$$f_k^x \to f \tag{6.29}$$

for some f in $L^2(G)$ as $k \to \infty$. Also, by (6.27), (6.28), Schwarz' inequality, and the fact that $\pi(g)$ is unitary for all g in G,

$$f_k^x(g) = (x, \pi(g)\varphi_k) \to (x, \pi(g)\varphi), \quad g \in G, \tag{6.30}$$

as $k \to \infty$. Thus, by (6.29) and (6.30), $(x, \pi(\cdot)\varphi) = f$ a.e. and hence

$$\int_G |(x, \pi(g)\varphi)|^2 d\mu(g) < \infty \tag{6.31}$$

for all x in X. Therefore, by (6.31),

$$\int_G |(\varphi, \pi(g)\varphi)|^2 d\mu(g) < \infty,$$

i.e., $\varphi \in \mathcal{D}$. Now, for any nonzero element x in X, we get, by (6.27), (6.29) and (6.30),

$$\begin{aligned}
\|\varphi_k - \varphi\|_q^2 &= \|\varphi_k - \varphi\|^2 + q(\varphi_k - \varphi, \varphi_k - \varphi) \\
&= \|\varphi_k - \varphi\|^2 + \frac{1}{\|x\|^2} \int_G |(x, \pi(g)(\varphi_k - \varphi))|^2 d\mu(g) \to 0
\end{aligned}$$

as $k \to \infty$. Hence $q : \mathcal{D} \times \mathcal{D} \to \mathbb{C}$ is closed. By Theorem 6.7, *i.e.*, the second representation theorem, there exists a positive and self-adjoint operator A from X into X with domain $\mathcal{D}$ such that

$$(A\varphi, A\psi) = q(\varphi, \psi), \quad \varphi, \psi \in \mathcal{D}. \tag{6.32}$$

Now, using (6.25), (6.26) and the unimodularity of G, the fact that $\mathcal{D}$ is an invariant subspace of X with respect to $\pi : G \to U(X)$ and the fact that $\pi : G \to U(X)$ is a representation, we get, for any nonzero element x in X,

$$
\begin{aligned}
q(\pi(h)\varphi, \pi(h)\psi) &= \frac{1}{\|x\|^2} \int_G (x, \pi(gh)\varphi)(\pi(gh)\psi, x) d\mu(g) \\
&= \frac{1}{\|x\|^2} \int_G (x, \pi(g)\varphi)(\pi(g)\psi, x) d\mu(g) \\
&= q(\varphi, \psi)
\end{aligned}
\tag{6.33}
$$

for all φ and ψ in $\mathcal{D}$, and all h in G. Thus, by (6.32), (6.33), and the fact that the representation $\pi : G \to U(X)$ is unitary,

$$
\begin{aligned}
(A\varphi, A\psi) &= (A\pi(h)\varphi, A\pi(h)\psi) \\
&= (\pi(h)\pi(h^{-1})A\pi(h)\varphi, A\pi(h)\psi) \\
&= (\pi(h^{-1})A\pi(h)\varphi, \pi(h^{-1})A\pi(h)\psi)
\end{aligned}
$$

for all φ and ψ in $\mathcal{D}$, and all h in G. Hence, on $\mathcal{D}$,

$$\pi(h^{-1})A\pi(h) = A, \quad h \in G,$$

or equivalently,

$$A\pi(h) = \pi(h)A, \quad h \in G.$$

Let $\mathcal{D}$ be the Hilbert space equipped with the inner product of which the induced norm is $\| \ \|_q$ given by (6.19). Then, by Theorem 6.4, *i.e.*, the extended Schur's lemma, A is a scalar multiple of an isometry from $\mathcal{D}$ into $L^2(G)$. So, we can find a positive number λ such that

$$\|A\varphi\|^2 = \lambda \|\varphi\|_q^2 = \lambda \|\varphi\|^2 + \lambda \|A\varphi\|^2, \quad \varphi \in \mathcal{D}. \tag{6.34}$$

Hence, by (6.34), $\lambda < 1$ and

$$\|A\varphi\|^2 = \frac{\lambda}{1 - \lambda} \|\varphi\|^2, \quad \varphi \in \mathcal{D}. \tag{6.35}$$

Using (6.35) and the density of $\mathcal{D}$ in X, we can extend $A : \mathcal{D} \to X$ to a bounded linear operator from X into X, which we denote by $\tilde{A} : X \to X$. So, if $\{\varphi_k\}_{k=1}^\infty$ is a sequence of elements in $\mathcal{D}$ such that $\varphi_k \to \varphi$ in X as $k \to \infty$, then $A\varphi_k \to \tilde{A}\varphi$ in $L^2(G)$ as $k \to \infty$. Since A is a closed linear operator from X into X with domain $\mathcal{D}$, it follows that $\varphi \in \mathcal{D}$. Therefore $\mathcal{D}$ is a closed subspace of X. Thus, using the irreducibility of $\pi : G \to U(X)$, we conclude that $\mathcal{D} = X$ and the proof is complete. $\qquad \square$

Remark 6.8 We give in Chapter 17 a unimodular group G, and an irreducible and unitary representation $\pi : G \to U(X)$ of G on X for which $AW(\pi) = \{x \in X : \|x\| = 1\}$. A different unimodular group G', and a new irreducible and unitary representation $\pi' : G' \to U(X)$ for which $AW(\pi) = \phi$ are also given. It is worth emphasizing the fact that Theorem 6.6 is false, in general, for non-unimodular groups, and Chapter 18 is devoted to a study of a non-unimodular group for which the conclusion of Theorem 6.6 is not true.

7 Wavelet Transforms

The linear operator $A_\varphi : X \to L^2(G)$ used in the proof of the resolution of the identity formula (6.3) in the previous chapter plays a pivotal role in this book. It is in fact the wavelet transform associated to the admissible wavelet φ for the irreducible and square-integrable representation $\pi : G \to U(X)$ of a locally compact and Hausdorff group G on a Hilbert space. To be more precise, we introduce the following definition.

Definition 7.1 Let φ be an admissible wavelet for a square-integrable representation $\pi : G \to U(X)$ of a locally compact and Hausdorff group G on a Hilbert space X. Then the wavelet transform associated to the admissible wavelet φ is the linear operator $A_\varphi : X \to C(G)$ defined by

$$(A_\varphi x)(g) = \frac{1}{\sqrt{c_\varphi}}(x, \pi(g)\varphi), \quad g \in G, \tag{7.1}$$

where $C(G)$ is the set of all continuous and complex-valued functions on G.

In the case when the representation $\pi : G \to U(X)$ in Definition 7.1 is also irreducible, the wavelet transform associated to φ is in fact equal to $\frac{1}{\sqrt{c_\varphi}}A_\varphi$, where A_φ is the linear operator used extensively in Chapter 6. To emphasize the fact that the wavelet transform is something we have come across before, we prefer to denote the wavelet transform associated to φ again by A_φ. There should be no danger of confusion. Another point to note is that the wavelet transform is defined for all square-integrable representations. The irreducibility condition is invoked only when it is necessary.

From the proof of Theorem 6.1, we see that if φ is an admissible wavelet for an irreducible and square-integrable representation $\pi : G \to U(X)$ of G on X, then $A_\varphi : X \to L^2(G)$ is a bounded linear operator. In this chapter and the following four chapters, we give a more detailed study of wavelet transforms.

We begin with the following reformulation of Theorem 6.1 in terms of the wavelet transform $A_\varphi : X \to L^2(G)$.

Theorem 7.2 *Let φ be an admissible wavelet for an irreducible and square-integrable representation $\pi : G \to U(X)$ of a locally compact and Hausdorff group G on a Hilbert space X. Then*

$$(x, y) = \int_G (A_\varphi x)(g)\overline{(A_\varphi y)(g)}d\mu(g)$$

for all x and y in X.

Remark 7.3 Theorem 7.2 can be considered as Plancherel's theorem for the wavelet transform $A_\varphi : X \to L^2(G)$.

Remark 7.4 That the wavelet transform $A_\varphi : X \to L^2(G)$ is an isometry is an immediate consequence of Theorem 7.2. Thus, the wavelet transform $A_\varphi : X \to L^2(G)$ is an isometry of X onto its range $R(A_\varphi)$. This is why the range $R(A_\varphi)$ of the wavelet transform $A_\varphi : X \to L^2(G)$ should be studied in some detail.

Proposition 7.5 *Let φ be an admissible wavelet for an irreducible and square-integrable representation of a locally compact and Hausdorff group G on a Hilbert space X. Then the range $R(A_\varphi)$ of $A_\varphi : X \to L^2(G)$ is a closed subspace of $L^2(G)$.*

Proof. Let $\{F_j\}_{j=1}^{\infty}$ be a sequence of functions in $R(A_\varphi)$ such that

$$F_j \to F \qquad (7.2)$$

for some F in $L^2(G)$ as $j \to \infty$. For $j = 1, 2, \ldots$, let x_j be such that $A_\varphi x_j = F_j$. Now, using Plancherel's theorem for $A_\varphi : X \to L^2(G)$,

$$\|x_j - x_k\| = \|A_\varphi x_j - A_\varphi x_k\|_{L^2(G)} = \|F_j - F_k\|_{L^2(G)} \to 0$$

as $j, k \to \infty$. Therefore $\{x_j\}_{j=1}^{\infty}$ is a Cauchy sequence in X. So, $x_j \to x$ for some x in X as $j \to \infty$. Thus,

$$F_j = A_\varphi x_j \to A_\varphi x \qquad (7.3)$$

in $L^2(G)$ as $j \to \infty$. By (7.2) and (7.3), $F = A_\varphi x$. Hence $F \in R(A_\varphi)$ and the proof is complete. $\qquad \square$

It follows from Proposition 7.5 that the range $R(A_\varphi)$ of the wavelet transform $A_\varphi : X \to L^2(G)$ is a Hilbert space. In fact, it is much more than a Hilbert space.

Theorem 7.6 *Let φ be an admissible wavelet for an irreducible and square-integrable representation of a locally compact and Hausdorff group G on a Hilbert space X. Then the range $R(A_\varphi)$ of the wavelet transform $A_\varphi : X \to L^2(G)$ is a reproducing kernel Hilbert space with reproducing kernel p_φ given by*

$$p_\varphi(g) = \frac{1}{c_\varphi}(\pi(g)\varphi, \varphi), \quad g \in G. \qquad (7.4)$$

In other words, if $F \in R(A_\varphi)$, then

$$F(g) = \int_G F(h)p_\varphi(g^{-1}h)d\mu(h), \quad g \in G.$$

Proof. Let $F \in R(A_\varphi)$. Then there exists an element x in X such that

$$F = A_\varphi x. \qquad (7.5)$$

Using the resolution of the identity formula in Chapter 6, (7.1), (7.4), (7.5) and the fact that $\pi : G \to U(X)$ is a unitary representation, we get

$$\int_G F(h)p_\varphi(g^{-1}h)d\mu(h)$$

$$= \frac{1}{c_\varphi}\int_G (A_\varphi x)(h)(\pi(g^{-1}h)\varphi, \varphi)d\mu(h)$$

$$= \frac{1}{c_\varphi^{3/2}}\int_G (x, \pi(h)\varphi)(\pi(h)\varphi, \pi(g)\varphi)d\mu(h)$$

$$= \frac{1}{\sqrt{c_\varphi}}(x, \pi(g)\varphi) = (A_\varphi x)(g) = F(g)$$

for all g in G.

$\square$

8   A Sampling Theorem

As an application of the fact that the range of the wavelet transform associated to an admissible wavelet for an irreducible and square-integrable representation is a reproducing kernel Hilbert space, we give in this chapter a sampling theorem on a locally compact and Hausdorff group. This is an analogue of Shannon's sampling theorem given in Section 2.4 of the book [7] by Blatter and Section 2.1 of the book [13] by Daubechies among others. The origin of the theorem is rooted in the papers [79, 80] by Shannon.

Let φ be an admissible wavelet for an irreducible and square-integrable representation $\pi : G \to U(X)$ of a locally compact and Hausdorff group G on a Hilbert space X. Let $\{g_j : j = 1, 2, \ldots\}$ be a countable collection of elements in G such that $\{\pi(g_j)\varphi : j = 1, 2, \ldots\}$ is an orthonormal basis for X. As has been indicated in Remark 7.3, $A_\varphi : X \to L^2(G)$ is an isometry. Thus, $\{A_\varphi \pi(g_j)\varphi : j = 1, 2, \ldots\}$ is an orthonormal basis for the range $R(A_\varphi)$ of $A_\varphi : X \to L^2(G)$. For $j = 1, 2, \ldots$, let $k_\varphi^{g_j}$ be the function on G defined by

$$k_\varphi^{g_j}(g) = (A_\varphi \pi(g_j)\varphi)(g), \quad g \in G. \tag{8.1}$$

Theorem 8.1 *Let $F \in R(A_\varphi)$. Then*

$$F = \sum_{j=1}^{\infty} (F, k_\varphi^{g_j})_{L^2(G)} k_\varphi^{g_j},$$

where the series is convergent in $L^2(G)$ and is absolutely convergent on G.

Proof. The convergence of the series in $L^2(G)$ follows from the fact that $\{k_\varphi^{g_j} : j = 1, 2, \ldots\}$ is an orthonormal basis for $R(A_\varphi)$. So, we only need to prove absolute convergence on G. Indeed, using Schwarz' inequality and Parseval's identity, we get

$$\sum_{j=1}^{\infty} |(F, k_\varphi^{g_j})_{L^2(G)}||k_\varphi^{g_j}(g)| \leq \left\{ \sum_{j=1}^{\infty} |(F, k_\varphi^{g_j})_{L^2(G)}|^2 \right\}^{\frac{1}{2}} \left\{ \sum_{j=1}^{\infty} |k_\varphi^{g_j}(g)|^2 \right\}^{\frac{1}{2}}$$

$$= \|F\|_{L^2(G)} \left\{ \sum_{j=1}^{\infty} |k_\varphi^{g_j}(g)|^2 \right\}^{\frac{1}{2}} \tag{8.2}$$

for all g in G. Since $k_\varphi^{g_j} \in R(A_\varphi)$, $j = 1, 2, \ldots$, and $R(A_\varphi)$ is a reproducing kernel Hilbert space, it follows that

$$k_\varphi^{g_j}(g) = \int_G k_\varphi^{g_j}(h) p_\varphi(g^{-1}h) d\mu(h) = (k_\varphi^{g_j}, \overline{p_\varphi^g})_{L^2(G)}, \quad g \in G, \qquad (8.3)$$

for $j = 1, 2, \ldots$, where p_φ^g is the function on G given by

$$p_\varphi^g(h) = p_\varphi(g^{-1}h), \quad h \in G. \qquad (8.4)$$

So, by (8.2)–(8.4) and Schwarz' inequality,

$$\sum_{j=1}^{\infty} |(F, k_\varphi^{g_j})_{L^2(G)}| \, |k_\varphi^{g_j}(g)| \leq \|F\|_{L^2(G)} \|p_\varphi^g\|_{L^2(G)}$$

and the proof is complete. $\qquad\qquad\qquad\qquad\qquad\qquad\qquad\qquad\qquad\quad$ □

The absolute convergence on G of the series in Theorem 8.1 says that

$$F(g) = \sum_{j=1}^{\infty} (F, k_\varphi^{g_j})_{L^2(G)} k_\varphi^{g_j}(g) = \sum_{j=1}^{\infty} \left\{ \int_G F(h) \overline{k_\varphi^{g_j}(h)} d\mu(h) \right\} k_\varphi^{g_j}(g) \qquad (8.5)$$

for all g in G. By (7.1), (7.4), (8.1) and the fact that $\pi : G \to U(X)$ is a unitary representation, we get for all h in G,

$$\begin{aligned} k_\varphi^{g_j}(h) &= (A_\varphi \pi(g_j)\varphi)(h) = \frac{1}{\sqrt{c_\varphi}} (\pi(g_j)\varphi, \pi(h)\varphi) \\ &= \frac{1}{\sqrt{c_\varphi}} (\varphi, \pi(g_j^{-1}h)\varphi) = \sqrt{c_\varphi} \overline{p_\varphi(g_j^{-1}h)} \end{aligned} \qquad (8.6)$$

for $j = 1, 2, \ldots$. So, by (8.5) and (8.6),

$$F(g) = \sqrt{c_\varphi} \sum_{j=1}^{\infty} \left\{ \int_G F(h) p_\varphi(g_j^{-1}h) d\mu(h) \right\} k_\varphi^{g_j}(g) = \sqrt{c_\varphi} \sum_{j=1}^{\infty} F(g_j) k_\varphi^{g_j}(g)$$

for all g in G. Thus,

$$F = \sqrt{c_\varphi} \sum_{j=1}^{\infty} F(g_j) k_\varphi^{g_j},$$

i.e., each signal F processed by means of the wavelet transform $A_\varphi : X \to L^2(G)$ can be reconstructed in terms of its sampled values $\{F(g_j) : j = 1, 2, \ldots\}$ on G. Therefore Theorem 8.1 is a sampling theorem.

9 Wavelet Constants

We begin with the following theorem, which is an extension of the resolution of the identity formula given in Theorem 6.1. This extension allows us to obtain some interesting results on the wavelet constants for unimodular groups.

Theorem 9.1 *Let φ and ψ be two admissible wavelets for an irreducible and square-integrable representation $\pi : G \to U(X)$ of a locally compact and Hausdorff group G on a Hilbert space X. Then for all x and y in X,*

$$\int_G (x, \pi(g)\varphi)(\pi(g)\psi, y)d\mu(g) = c_{\varphi,\psi}(x, y), \tag{9.1}$$

where

$$c_{\varphi,\psi} = \int_G (\varphi, \pi(g)\varphi)(\pi(g)\psi, \varphi)d\mu(g). \tag{9.2}$$

Proof. Using the definition of the wavelet transform and the fact that $\pi : G \to U(X)$ is a unitary representation, we get

$$(A_\varphi \pi(h)x)(g) = \frac{1}{\sqrt{c_\varphi}}(\pi(h)x, \pi(g)\varphi) = \frac{1}{\sqrt{c_\varphi}}(x, \pi(h^{-1}g)\varphi) = (A_\varphi x)(h^{-1}g), \tag{9.3}$$

for all x in X, and all g and h in G. So,

$$A_\varphi \pi(h) = L(h)A_\varphi, \tag{9.4}$$

where $L : G \to U(L^2(G))$ is the left regular representation of G. Now, for all x and y in X, we get, by (9.3), (9.4), the left invariance of μ and the fact that $\pi : G \to U(X)$ is a unitary representation,

$$
\begin{aligned}
(A_\psi^* L(g)A_\varphi x, y) &= (L_g A_\varphi x, A_\psi y)_{L^2(G)} \\
&= \int_G (A_\varphi x)(g^{-1}h)\overline{(A_\psi y)(h)}d\mu(h) \\
&= \int_G (A_\varphi x)(h)\overline{(A_\psi y)(gh)}d\mu(h) \\
&= (A_\varphi x, L_{g^{-1}}A_\psi y)_{L^2(G)} \\
&= (A_\varphi x, A_\psi \pi(g^{-1})y)_{L^2(G)} \\
&= (\pi(g)A_\psi^* A_\varphi x, y)_{L^2(G)}, \quad g \in G, \tag{9.5}
\end{aligned}
$$

where $A_\psi^* : L^2(G) \to X$ is the adjoint of $A_\psi : X \to L^2(G)$. Thus, by (9.5), we get

$$A_\psi^* L(g) A_\varphi = \pi(g) A_\psi^* A_\varphi, \quad g \in G. \tag{9.6}$$

Hence, by (9.4) and (9.6), we get

$$A_\psi^* A_\varphi \pi(g) = \pi(g) A_\psi^* A_\varphi, \quad g \in G. \tag{9.7}$$

From (9.7) and the fact that $\pi : G \to U(X)$ is an irreducible representation, we can use Schur's lemma to conclude that there exists a constant $\alpha_{\varphi,\psi}$ such that

$$A_\psi^* A_\varphi = \alpha_{\varphi,\psi} I, \tag{9.8}$$

where I is the identity operator on X. So, by (9.8), we get

$$
\begin{aligned}
\alpha_{\varphi,\psi}(x,y) &= (A_\psi^* A_\varphi x, y) = (A_\varphi x, A_\psi y)_{L^2(G)} \\
&= \frac{1}{\sqrt{c_\varphi c_\psi}} \int_G (x, \pi(g)\varphi)(\pi(g)\psi, y) d\mu(g)
\end{aligned}
\tag{9.9}
$$

for all x and y in X. If we let $x = y = \varphi$ in (9.9) and use (9.2), then we get

$$\alpha_{\varphi,\psi} = \frac{c_{\varphi,\psi}}{\sqrt{c_\varphi c_\psi}}$$

and (9.1) follows. $\square$

We call the number $c_{\varphi,\psi}$ in Theorem 9.1 the two-wavelet constant associated to the admissible wavelets φ and ψ. It is obvious that $c_{\varphi,\varphi}$ is the same as the wavelet constant c_φ associated to the admissible wavelet φ defined by (6.2). As an application of two-wavelet constants, we give the following immediate consequence of Theorem 9.1.

Corollary 9.2 *Let φ and ψ be two admissible wavelets for an irreducible and square-integrable representation of a locally compact and Hausdorff group G on a Hilbert space X. If $c_{\varphi,\psi} = 0$, then $R(A_\varphi)$ and $R(A_\psi)$ are orthogonal.*

For unimodular groups, the wavelet constants are particularly illuminating. The following theorem gives us very important information on two-wavelet constants.

Theorem 9.3 *Let G be a unimodular group, and let φ and ψ be two admissible wavelets for an irreducible and square-integrable representation $\pi : G \to U(X)$ of G on a Hilbert space X. Then*

$$c_{\varphi,\psi} = (\psi, \varphi) c_\varphi. \tag{9.10}$$

Proof. By (9.2) and the fact that $\pi : G \to U(X)$ is a unitary representation,

$$
\begin{aligned}
c_{\varphi,\psi} &= \int_G (\varphi, \pi(g)\varphi)(\pi(g)\psi, \varphi)d\mu(g) \\
&= \int_G (\psi, \pi(g^{-1})\varphi)(\pi(g^{-1})\varphi, \varphi)d\mu(g) \\
&= \int_G (\psi, \pi(g)\varphi)(\pi(g)\varphi, \varphi)d\tilde{\mu}(g), \quad\quad (9.11)
\end{aligned}
$$

where $\tilde{\mu}$ is defined by (4.5). So, using the unimodularity of G, Theorem 6.1 and (9.11), we get (9.10). $\qquad\square$

The following consequence of (9.2) and Theorem 9.3 holds for unimodular groups, and is an interesting result in its own right.

Theorem 9.4 *Let G be a unimodular group, and let φ and ψ be two admissible wavelets for an irreducible and square-integrable representation $\pi : G \to U(X)$ of G on a Hilbert space X. Then $c_\varphi = c_\psi$.*

Proof. By putting $x = y = \psi$ in (9.1), we get

$$
c_{\varphi,\psi} = \int_G (\psi, \pi(g)\varphi)(\pi(g)\psi, \psi)d\mu(g). \quad\quad (9.12)
$$

By (9.2) and (9.12), we get
$$
c_{\psi,\varphi} = \overline{c_{\varphi,\psi}}. \quad\quad (9.13)
$$
So, by Theorem 9.3, (9.13) and the fact that c_φ is real-valued,

$$
(\varphi, \psi)c_\varphi = \overline{(\psi, \varphi)c_\varphi} = \overline{c_{\varphi,\psi}} = c_{\psi,\varphi} = (\varphi, \psi)c_\psi.
$$

So,
$$
c_\varphi = c_\psi
$$
if $(\varphi, \psi) \neq 0$. Now, suppose that $(\varphi, \psi) = 0$. Let $\omega \in X$ be such that $\|\omega\| = 1$, $(\omega, \varphi) \neq 0$ and $(\omega, \psi) \neq 0$. By Theorem 6.6, ω is also an admissible wavelet for the square-integrable representation $\pi : G \to U(X)$. Thus, by what we have just shown, $c_\varphi = c_\omega$ and $c_\psi = c_\omega$. Hence $c_\varphi = c_\psi$ and the proof is complete. $\qquad\square$

For unimodular groups, the following theorem gives us a decomposition of a wavelet transform into an orthogonal direct sum of wavelet transforms. It amounts to saying that any signal F on G, which comes as a wavelet transform, is a superposition of signals obtained as wavelet transforms associated to admissible wavelets which form an orthonormal basis for X.

Theorem 9.5 *Let $\{\varphi_j : j = 1, 2, \ldots\}$ be an orthonormal basis for X. Then for all φ in X with $\|\varphi\| = 1$,*

$$
A_\varphi x = \sum_{j=1}^{\infty} \oplus \overline{(\varphi, \varphi_j)} A_{\varphi_j} x, \quad x \in X.
$$

Proof. For all x in X and all g in G, we get

$$
\begin{aligned}
(A_\varphi x)(g) &= \frac{1}{\sqrt{c_\varphi}}(x, \pi(g)\varphi) \\
&= \frac{1}{\sqrt{c_\varphi}}\left(x, \pi(g)\sum_{j=1}^{\infty}(\varphi, \varphi_j)\varphi_j\right) \\
&= \frac{1}{\sqrt{c_\varphi}}\sum_{j=1}^{\infty}\overline{(\varphi, \varphi_j)}(x, \pi(g)\varphi_j) \\
&= \sum_{j=1}^{\infty}\overline{(\varphi, \varphi_j)}(A_{\varphi_j} x)(g).
\end{aligned} \tag{9.14}
$$

Thus, by (9.14), Theorems 9.1 and 9.3, the proof is complete. $\square$

10 Adjoints

The left regular representation $L : G \to U(L^2(G))$ of a locally compact and Hausdorff group G given by

$$(L(g)u)(h) = u(g^{-1}h), \quad g, h \in G,$$

for all u in $L^2(G)$, has been playing a subsidiary role since its appearance in Example 5.7. In this chapter we show that it is an object of interest in its own right. We are particularly interested in the adjoints of wavelet transforms for left regular representations of unimodular groups.

In order to understand why the wavelet transform associated to the left regular representation of a group is important, let us recall that a filter in electrical engineering is a bounded linear operator from $L^2(G)$ into $L^2(G)$, which commutes with $L(g)$ for all g in G. That the left translation invariance should be an indispensable property of a filter is explained in, *e.g.*, the book [29] by Gasquet and Witomsky.

Theorem 10.1 *Let φ be an admissible wavelet for the left regular representation $L : G \to U(L^2(G))$ of G. Then the wavelet transform $A_\varphi : L^2(G) \to L^2(G)$ is a filter.*

Proof. Using the definition of the wavelet transform $A_\varphi : L^2(G) \to L^2(G)$, we get

$$
\begin{aligned}
(A_\varphi L(h)u)(g) &= \frac{1}{\sqrt{c_\varphi}}(L(h)u, L(g)\varphi)_{L^2(G)} \\
&= \frac{1}{\sqrt{c_\varphi}}(u, L(h^{-1}g)\varphi)_{L^2(G)} \\
&= (A_\varphi u)(h^{-1}g) = (L(h)A_\varphi u)(g), \quad g, h \in G,
\end{aligned}
$$

for all u in $L^2(G)$, and the proof is complete. □

We have the following result, which tells us that if G is unimodular, then admissible wavelets for the left regular representation of G abound.

Theorem 10.2 *Let $L : G \to U(L^2(G))$ be the left regular representation of a unimodular group G. Then every function φ in $L^1(G) \cap L^2(G)$ with $\|\varphi\|_{L^2(G)} = 1$ is an admissible wavelet for $L : G \to U(L^2(G))$.*

Proof. Using Minkowski's inequality in integral form, the unimodularity of the group G and $\|\varphi\|_{L^2(G)} = 1$, we get

$$\left\{ \int_G |(\varphi, L(g)\varphi)|^2 d\mu(g) \right\}^{\frac{1}{2}} = \left\{ \int_G \left| \int_G \varphi(h)\overline{\varphi(g^{-1}h)}d\mu(h) \right|^2 d\mu(g) \right\}^{\frac{1}{2}}$$

$$\leq \int_G \left\{ \int_G |\varphi(h)|^2 |\varphi(g^{-1}h)|^2 d\mu(g) \right\}^{\frac{1}{2}} d\mu(h)$$

$$= \int_G |\varphi(h)| \left\{ \int_G |\varphi(g^{-1}h)|^2 d\mu(g) \right\}^{\frac{1}{2}} d\mu(h)$$

$$= \|\varphi\|_{L^1(G)} \|\varphi\|_{L^2(G)} = \|\varphi\|_{L^1(G)} < \infty,$$

and the proof is complete. □

In order to compute the adjoint of a wavelet transform associated to an admissible wavelet for a left regular representation of a unimodular group, we need the following lemma.

Lemma 10.3 *Let $\varphi \in L^1(G) \cap L^2(G)$ be such that $\|\varphi\|_{L^2(G)} = 1$, where G is a unimodular group. Then so is the function φ^* on G defined by*

$$\varphi^*(g) = \overline{\varphi(g^{-1})}, \quad g \in G.$$

The proof of Lemma 10.3 is so simple that we can omit it.

We can now give a formula for the adjoint of a wavelet transform associated to an admissible wavelet for a left regular representation of a unimodular group.

Theorem 10.4 *Let $\varphi \in L^1(G) \cap L^2(G)$ be such that $\|\varphi\|_{L^2(G)} = 1$, where G is a unimodular group. Then for the left regular representation $L : G \to U(L^2(G))$ of G, the adjoint of the wavelet transform $A_\varphi : L^2(G) \to L^2(G)$ is equal to $\sqrt{\frac{c_{\varphi^*}}{c_\varphi}} A_{\varphi^*} : L^2(G) \to L^2(G)$.*

Proof. Let u and v be in $L^2(G)$. Then, using the unimodularity of G, Young's inequality and Fubini's theorem, we get

$$(A_\varphi u, v)_{L^2(G)} = \int_G (A_\varphi u)(g)\overline{v(g)}d\mu(g)$$

$$= \frac{1}{\sqrt{c_\varphi}} \int_G \left\{ \int_G u(h)\overline{(L(g)\varphi)(h)}d\mu(h) \right\} \overline{v(g)}d\mu(g)$$

$$= \frac{1}{\sqrt{c_\varphi}} \int_G u(h) \overline{\left\{ \int_G (L(g)\varphi)(h)v(g)d\mu(g) \right\}} d\mu(h)$$

$$= \frac{1}{\sqrt{c_\varphi}} \int_G u(h) \overline{\left\{ \int_G \varphi(g^{-1}h)v(g)d\mu(g) \right\}} d\mu(h)$$

$$= \frac{1}{\sqrt{c_\varphi}} \int_G u(h) \left\{ \overline{\int_G v(g)\overline{\varphi^*(h^{-1}g)}d\mu(g)} \right\} d\mu(h)$$

$$= \sqrt{\frac{c_{\varphi^*}}{c_\varphi}} \int_G u(h)\overline{(A_{\varphi^*}v)(h)}d\mu(h) = \sqrt{\frac{c_{\varphi^*}}{c_\varphi}}(u, A_{\varphi^*}v)_{L^2(G)},$$

and the proof is complete. □

An immediate corollary of Theorems 9.4 and 10.4 is the following result.

Corollary 10.5 *If the left regular representation $L : G \to U(L^2(G))$ of a unimodular group G is irreducible, then for all φ in $L^1(G) \cap L^2(G)$ with $\|\varphi\|_{L^2(G)} = 1$, the adjoint of the wavelet transform $A_\varphi : L^2(G) \to L^2(G)$ is the wavelet transform $A_{\varphi^*} : L^2(G) \to L^2(G)$.*

An important corollary of Theorem 10.4 is the following characterization of self-adjoint wavelet transforms for left regular representations of unimodular groups.

Theorem 10.6 *Let $\varphi \in L^1(G) \cap L^2(G)$ be such that $\|\varphi\|_{L^2(G)} = 1$, where G is a unimodular group. Then for the left regular representation $L : G \to U(L^2(G))$ of G, the wavelet transform $A_\varphi : L^2(G) \to L^2(G)$ is self-adjoint if and only if $\varphi = \varphi^*$.*

Theorem 10.6 follows immediately from Theorem 10.4 and the following lemma.

Lemma 10.7 *Let φ and ψ be two admissible wavelets associated to the left regular representation $L : G \to U(L^2(G))$ of a locally compact and Hausdorff group G such that $c_\varphi = c_\psi$ and $A_\varphi = A_\psi$. Then $\varphi = \psi$.*

Proof. Using the definition of the wavelet transform and the fact that $L : G \to U(L^2(G))$ is a unitary representation of G, we get

$$(L(g)u, \varphi)_{L^2(G)} = (L(g)u, \psi)_{L^2(G)}, \quad g \in G,$$

for all u in $L^2(G)$. Thus, if we let g be the identity element in the group G, then we get $\varphi = \psi$. □

Remark 10.8 We end this chapter with the observation that if $\varphi \in L^1(G) \cap L^2(G)$ is such that $\|\varphi\|_{L^2(G)} = 1$, where G is a unimodular group, then for the left regular representation $L : G \to U(L^2(G))$ of G, the wavelet transform $A_\varphi : L^2(G) \to L^2(G)$ associated to φ is in fact the convolution operator from $L^2(G)$ into $L^2(G)$ of which the kernel is the function $\frac{1}{\sqrt{c_\varphi}}\overline{\varphi}$. Indeed,

$$(A_\varphi u)(g) = \frac{1}{\sqrt{c_\varphi}}(u, L(g)\varphi)_{L^2(G)} = \frac{1}{\sqrt{c_\varphi}}\int_G u(h)\overline{\varphi(g^{-1}h)}d\mu(h), \quad g \in G,$$

for all u in $L^2(G)$.

11 Compact Groups

We look at left regular representations $L : G \to U(L^2(G))$ of compact and Hausdorff groups G in this chapter. Let $\varphi \in L^2(G)$. Then, using Minkowski's inequality in integral form, the unimodularity of the group G and Schwarz' inequality, we get

$$
\left\{ \int_G |(\varphi, L(g)\varphi)|^2 d\mu(g) \right\}^{\frac{1}{2}} = \left\{ \int_G \left| \int_G \varphi(h)\overline{\varphi(g^{-1}h)} d\mu(h) \right|^2 d\mu(g) \right\}^{\frac{1}{2}}
$$

$$
\leq \int_G \left\{ \int_G |\varphi(h)|^2 |\varphi(g^{-1}h)|^2 d\mu(g) \right\}^{\frac{1}{2}} d\mu(h)
$$

$$
= \int_G |\varphi(h)| \left\{ \int_G |\varphi(g^{-1}h)|^2 d\mu(g) \right\}^{\frac{1}{2}} d\mu(h)
$$

$$
= \|\varphi\|_{L^1(G)} \|\varphi\|_{L^2(G)} \leq \mu(G)^{\frac{1}{2}} \|\varphi\|_{L^2(G)}^2 < \infty.
$$

Thus, every function φ in $L^2(G)$ with $\|\varphi\|_{L^2(G)} = 1$ is an admissible wavelet for the left regular representation $L : G \to U(L^2(G))$ of G.

Theorem 11.1 *Let $\varphi \in L^2(G)$ be such that $\|\varphi\|_{L^2(G)} = 1$, where G is a compact and Hausdorff group. Then for the left regular representation $L : G \to U(L^2(G))$ of G, the wavelet transform $A_\varphi : L^2(G) \to L^2(G)$ is in the Hilbert-Schmidt class S_2 and*

$$
\|A_\varphi\|_{S_2} = \frac{1}{\sqrt{c_\varphi}} (\mu(G))^{\frac{1}{2}}.
$$

Proof. Let $\{\varphi_k : k = 1, 2, \ldots\}$ be an orthonormal basis for $L^2(G)$. Then, using Fubini's theorem, Parseval's identity, the fact that $L(g) : L^2(G) \to L^2(G)$ is a unitary operator for all g in G, and the fact that $\|\varphi\|_{L^2(G)} = 1$, we get

$$
\sum_{k=1}^{\infty} \|A_\varphi \varphi_k\|_{L^2(G)}^2 = \frac{1}{c_\varphi} \sum_{k=1}^{\infty} \int_G |(\varphi_k, \pi(g)\varphi)_{L^2(G)}|^2 d\mu(g)
$$

$$
= \frac{1}{c_\varphi} \int_G \sum_{k=1}^{\infty} |(\varphi_k, \pi(g)\varphi)_{L^2(G)}|^2 d\mu(g)
$$

$$
= \frac{1}{c_\varphi} \int_G \|\pi(g)\varphi\|_{L^2(G)}^2 d\mu(g) = \frac{1}{c_\varphi} \mu(G),
$$

and hence, by Proposition 2.8, the proof is complete. $\square$

   The significance of the Hilbert-Schmidt property of the wavelet transform $A_\varphi : L^2(G) \to L^2(G)$ is revealed in the following theorem, which tells us how fast the wavelet transform $A_\varphi : L^2(G) \to L^2(G)$ can be approximated by the sequence of finite truncations of its singular value decomposition.

Theorem 11.2 *Let $\varphi \in L^2(G)$ be such that $\|\varphi\|_{L^2(G)} = 1$, where G is a compact and Hausdorff group. Then for the left regular representation $L : G \to U(L^2(G))$ of G, let*

$$A_\varphi u = \sum_{k=1}^{\infty} s_k(A_\varphi)(u, u_k)_{L^2(G)} v_k, \quad u \in L^2(G),$$

be the singular value decomposition of the wavelet transform $A_\varphi : L^2(G) \to L^2(G)$, where $s_k(A_\varphi)$, $k = 1, 2, \ldots$, are the singular values of the wavelet transform $A_\varphi : L^2(G) \to L^2(G)$, $\{u_k : k = 1, 2, \ldots\}$ is an orthonormal basis for $L^2(G)$, $\{v_k : k = 1, 2, \ldots\}$ is an orthonormal set for $L^2(G)$, and the convergence of the series is understood to be in $L^2(G)$. For $N = 1, 2, \ldots$, let $A_N : L^2(G) \to L^2(G)$ be the Nth truncation of the singular value decomposition given by

$$A_N u = \sum_{k=1}^{N} s_k(A_\varphi)(u, u_k)_{L^2(G)} v_k, \quad u \in L^2(G).$$

Then

$$\|A_N - A_\varphi\|_{B(L^2(G))} \leq \left\{ \sum_{k=N+1}^{\infty} (s_k(A_\varphi))^2 \right\}^{\frac{1}{2}} \to 0$$

as $N \to \infty$, where $\| \ \|_{B(L^2(G))}$ is the norm in the C^*-algebra of all bounded linear operators from $L^2(G)$ into $L^2(G)$.

Proof. We begin with the fact that

$$\|A\|_* \leq \|A\|_{S_2} \tag{11.1}$$

for all bounded linear operators $A : X \to X$ in S_2. Indeed, for all A in S_2, the canonical form for compact operators given by Theorem 2.2 says that

$$Ax = \sum_{j=1}^{\infty} s_j(A)(x, u_j) v_j, \quad x \in X, \tag{11.2}$$

where $s_j(A)$, $j = 1, 2, \ldots$, are the positive singular values of $A : X \to X$, $\{u_j : j = 1, 2, \ldots\}$ is an orthonormal basis for $N(A)^\perp$, $\{v_j : j = 1, 2, \ldots\}$ is an orthonormal set in X, and the convergence of the series is understood to be in X. Thus, by (11.2), the orthonormality of $\{v_j : j = 1, 2, \ldots\}$, the normality of $\{u_j : j = 1, 2, \ldots\}$

and the definition of $\| \; \|_{S_2}$,

$$
\begin{aligned}
\|Ax\|^2 &= \sum_{j,k=1}^{\infty} s_j(A)s_k(A)(x,u_j)\overline{(x,u_k)}(v_j,v_k) \\
&= \sum_{j=1}^{\infty}(s_j(A))^2|(x,u_j)|^2 \le \|A\|_{S_2}^2\|x\|^2, \quad x \in X,
\end{aligned}
$$

and hence (11.1) follows. Now, by (11.1), we get

$$
\begin{aligned}
\|A_N - A_\varphi\|_{B(L^2(G))}^2 &\le \|A_N - A_\varphi\|_{S_2}^2 = \sum_{k=N+1}^{\infty}\|A_\varphi u_k\|_{L^2(G)}^2 \\
&= \sum_{k=N+1}^{\infty}\|s_k(A_\varphi)u_k\|_{L^2(G)}^2 \le \sum_{k=N+1}^{\infty}(s_k(A_\varphi))^2 \to 0
\end{aligned}
$$

as $N \to \infty$. $\square$

It is interesting to note that Theorem 11.2 tells us that if G is a compact and Hausdorff group, then for the left regular representation $L : G \to U(L^2(G))$ of G, the wavelet transform $A_\varphi : L^2(G) \to L^2(G)$ associated to any function φ in $L^2(G)$ with $\|\varphi\|_{L^2(G)} = 1$ is compact and hence its range cannot be a closed subspace of $L^2(G)$. Thus, the wavelet transform $A_\varphi : L^2(G) \to L^2(G)$ cannot be an isometry. This is in sharp contrast with the theory of wavelet transforms associated to irreducible and square-integrable representations given in Chapter 7.

12 Localization Operators

Let φ be an admissible wavelet for an irreducible and square-integrable representation $\pi : G \to U(X)$ of a locally compact and Hausdorff group G on a Hilbert space X. In this chapter we introduce a class of bounded linear operators $L_{F,\varphi} : X \to X$, which are related to the wavelet transform $A_\varphi : X \to L^2(G)$ defined by (7.1), for all F in $L^p(G)$, $1 \le p \le \infty$. We first tackle this problem for F in $L^1(G)$ or $L^\infty(G)$. In the case when $p = 1$, we do not need the assumption that the representation $\pi : G \to U(X)$ be irreducible.

Let $F \in L^1(G) \cup L^\infty(G)$. Then for all x in X, we define $L_{F,\varphi}x$ by

$$(L_{F,\varphi}x, y) = \frac{1}{c_\varphi} \int_G F(g)(x, \pi(g)\varphi)(\pi(g)\varphi, y)d\mu(g) \tag{12.1}$$

for all y in X. Then we have the following proposition.

Proposition 12.1 *Let $F \in L^1(G)$. Then $L_{F,\varphi} : X \to X$ is a bounded linear operator and*

$$\|L_{F,\varphi}\|_* \le \frac{1}{c_\varphi}\|F\|_{L^1(G)}.$$

Proof. Let x and y be elements in X. Then, using (12.1), Schwarz' inequality, $\|\varphi\| = 1$ and the fact that $\pi(g) : X \to X$ is unitary for all g in G, we have

$$|(x, \pi(g)\varphi)(\pi(g)\varphi, y)| \le \|x\|\,\|y\|. \tag{12.2}$$

Since $F \in L^1(G)$, it follows from (12.1) and (12.2) that

$$|(L_{F,\varphi}x, y)| \le \frac{1}{c_\varphi}\|F\|_{L^1(G)}\|x\|\,\|y\|$$

and the proof of the proposition is complete. $\qquad\square$

We also have the following proposition.

Proposition 12.2 *Let $F \in L^\infty(G)$. Then $L_{F,\varphi} : X \to X$ is a bounded linear operator and*

$$\|L_{F,\varphi}\|_* \le \|F\|_{L^\infty(G)}.$$

Proof. Let x and y be elements in X. Then, using the resolution of the identity formula in Theorem 6.1, we get

$$\|x\|^2 = \frac{1}{c_\varphi} \int_G |(x, \pi(g)\varphi)|^2 d\mu(g) \tag{12.3}$$

and

$$\|y\|^2 = \frac{1}{c_\varphi} \int_G |(\pi(g)\varphi, y)|^2 d\mu(g). \tag{12.4}$$

Thus, using (12.3), (12.4), Schwarz' inequality, and the assumption that $F \in L^\infty(G)$, we get

$$|(L_{F,\varphi}x, y)| \leq \frac{1}{c_\varphi} \|F\|_{L^\infty(G)} \|x\| \|y\|,$$

and this completes the proof of the proposition. $\square$

We can now associate a localization operator $L_{F,\varphi} : X \to X$ to every function F in $L^p(G)$, $1 < p < \infty$, and prove that $L_{F,\varphi} : X \to X$ is a bounded linear operator. The precise result is the following proposition.

Proposition 12.3 *Let $F \in L^p(G)$, $1 < p < \infty$. Then there exists a unique bounded linear operator $L_{F,\varphi} : X \to X$ such that*

$$\|L_{F,\varphi}\|_* \leq \left(\frac{1}{c_\varphi}\right)^{\frac{1}{p}} \|F\|_{L^p(G)}, \tag{12.5}$$

and $L_{F,\varphi}x$ is given by (12.1) for all x in X and all simple functions F on G for which

$$\mu\{g \in G : F(g) \neq 0\} < \infty.$$

To prove Proposition 12.3, we need a recall of the Riesz-Thorin theorem given in, *e.g.*, Chapter 10 of the book [102] by Wong. It is in fact a special case of Theorem 2.10 in Chapter 2.

Theorem 12.4 (The Riesz-Thorin Theorem) *Let (X, μ) be a measure space and (Y, ν) a σ-finite measure space. Let T be a linear transformation with domain $\mathcal{D}$ consisting of all simple functions f on X such that*

$$\mu\{s \in X : f(s) \neq 0\} < \infty$$

and such that the range of T is contained in the set of all measurable functions on Y. Suppose that $\alpha_1, \alpha_2, \beta_1$ and β_2 are numbers in $[0, 1]$ and there exist positive constants M_1 and M_2 such that

$$\|Tf\|_{L^{\frac{1}{\beta_j}}(Y)} \leq M_j \|f\|_{L^{\frac{1}{\alpha_j}}(X)}, \quad f \in \mathcal{D}, \ j = 1, 2.$$

Then for $0 < \theta < 1$,

$$\alpha = (1-\theta)\alpha_1 + \theta\alpha_2 \quad and \quad \beta = (1-\theta)\beta_1 + \theta\beta_2,$$

we have

$$\|Tf\|_{L^{\frac{1}{\beta}}(Y)} \leq M_1^{1-\theta} M_2^\theta \|f\|_{L^{\frac{1}{\alpha}}(X)}, \quad f \in \mathcal{D}.$$

Proof of Proposition 12.3. Let $U : X \to L^2(\mathbb{R}^n)$ be a unitary operator between X and $L^2(\mathbb{R}^n)$. Let $F \in L^1(G)$. Then, by Proposition 12.1, the linear operator $\tilde{L}_{F,\varphi} : L^2(\mathbb{R}^n) \to L^2(\mathbb{R}^n)$, defined by

$$\tilde{L}_{F,\varphi} = U L_{F,\varphi} U^{-1}, \tag{12.6}$$

is bounded and

$$\|\tilde{L}_{F,\varphi}\|_{B(L^2(\mathbb{R}^n))} \leq \frac{1}{c_\varphi} \|F\|_{L^1(G)}, \tag{12.7}$$

where $\|\ \|_{B(L^2(\mathbb{R}^n))}$ is the norm in the C^*-algebra $B(L^2(\mathbb{R}^n))$ of all bounded linear operators from $L^2(\mathbb{R}^n)$ into $L^2(\mathbb{R}^n)$. If $F \in L^\infty(G)$, then, by Proposition 12.2, the linear operator $\tilde{L}_{F,\varphi} : L^2(\mathbb{R}^n) \to L^2(\mathbb{R}^n)$, defined by (12.6), is also bounded and

$$\|\tilde{L}_{F,\varphi}\|_{B(L^2(\mathbb{R}^n))} \leq \|F\|_{L^\infty(G)}. \tag{12.8}$$

Let $\mathcal{D}$ be the set of all simple functions F on G such that

$$\mu\{g \in G : F(g) \neq 0\} < \infty.$$

Let $f \in L^2(\mathbb{R}^n)$ and T be the linear transformation from $\mathcal{D}$ into the set of all Lebesgue measurable functions on $\mathbb{R}^n$ defined by

$$TF = \tilde{L}_{F,\varphi} f, \quad F \in \mathcal{D}. \tag{12.9}$$

Then, by (12.7) and (12.8),

$$\|TF\|_{L^2(\mathbb{R}^n)} \leq \frac{1}{c_\varphi} \|F\|_{L^1(G)} \|f\|_{L^2(\mathbb{R}^n)}$$

and

$$\|TF\|_{L^2(\mathbb{R}^n)} \leq \|F\|_{L^\infty(G)} \|f\|_{L^2(\mathbb{R}^n)}$$

for all functions F in $\mathcal{D}$. Thus, by Theorem 12.4,

$$\|TF\|_{L^2(\mathbb{R}^n)} \leq \left(\frac{1}{c_\varphi}\right)^{\frac{1}{p}} \|F\|_{L^p(G)} \|f\|_{L^2(\mathbb{R}^n)}, \quad F \in \mathcal{D}. \tag{12.10}$$

Therefore, by (12.9) and (12.10),

$$\|\tilde{L}_{F,\varphi} f\|_{L^2(\mathbb{R}^n)} \leq \left(\frac{1}{c_\varphi}\right)^{\frac{1}{p}} \|F\|_{L^p(G)} \|f\|_{L^2(\mathbb{R}^n)}, \quad F \in \mathcal{D}. \tag{12.11}$$

Since (12.11) is true for arbitrary functions f in $L^2(\mathbb{R}^n)$, it follows that

$$\|\tilde{L}_{F,\varphi}\|_{B(L^2(\mathbb{R}^n))} \leq \left(\frac{1}{c_\varphi}\right)^{\frac{1}{p}} \|F\|_{L^p(G)}, \quad F \in \mathcal{D}. \tag{12.12}$$

Let $F \in L^p(G)$, $1 < p < \infty$. Then there exists a sequence $\{F_k\}_{k=1}^\infty$ of functions in $\mathcal{D}$ such that $F_k \to F$ in $L^p(G)$ as $k \to \infty$. By (12.12),

$$\|\tilde{L}_{F_j,\varphi} - \tilde{L}_{F_k,\varphi}\|_{B(L^2(\mathbb{R}^n))} \le \left(\frac{1}{c_\varphi}\right)^{\frac{1}{p}} \|F_j - F_k\|_{L^p(G)} \to 0$$

as $j, k \to \infty$. Therefore $\{\tilde{L}_{F_k,\varphi}\}_{k=1}^\infty$ is a Cauchy sequence in $B(L^2(\mathbb{R}^n))$. Using the completeness of $B(L^2(\mathbb{R}^n))$, we can find a bounded linear operator $\tilde{L}_{F,\varphi} : L^2(\mathbb{R}^n) \to L^2(\mathbb{R}^n)$ such that $\tilde{L}_{F_k,\varphi} \to \tilde{L}_{F,\varphi}$ as $k \to \infty$. Since each $\tilde{L}_{F_k,\varphi}$ satisfies (12.12), it follows that $\tilde{L}_{F,\varphi}$ also satisfies (12.12). Thus, the linear operator $L_{F,\varphi} : X \to X$, where

$$L_{F,\varphi} = U^{-1}\tilde{L}_{F,\varphi}U,$$

is a bounded linear operator satisfying the conclusions of the theorem if $F \in L^p(G)$, $1 < p < \infty$. To prove uniqueness, let $F \in L^p(G)$, $1 < p < \infty$, and suppose that $P_F : X \to X$ is another bounded linear operator satisfying the conclusions of the theorem. Let $Q : L^p(G) \to B(X)$ be the linear operator defined by

$$QF = L_{F,\varphi} - P_F, \quad F \in L^p(G).$$

Then, by (12.6),

$$\|QF\|_* \le 2\left(\frac{1}{c_\varphi}\right)^{\frac{1}{p}} \|F\|_{L^p(G)}, \quad F \in L^p(G).$$

Furthermore, QF is equal to the zero operator on X for all F in $\mathcal{D}$. Thus, $Q : L^p(G) \to B(X)$ is a bounded linear operator that is equal to zero on the dense subspace $\mathcal{D}$ of $L^p(G)$. Therefore $P_F = L_{F,\varphi}$ for all functions F in $L^p(G)$. $\square$

Remark 12.5 The bounded linear operators $L_{F,\varphi} : X \to X$ introduced in this chapter are dubbed localization operators in the paper [40]. The impetus for the terminology stems from the simple observation that if $F(g) = 1$ for all g in G, then the resolution of the identity formula in Theorem 6.1 implies that the corresponding linear operator is simply the identity operator on X. Thus, in general, the function F, which is also called the symbol of the localization operator $L_{F,\varphi} : X \to X$, is there to localize on G so as to produce a nontrivial bounded linear operator on X with various applications in the mathematical sciences.

13 S_p Norm Inequalities, $1 \leq p \leq \infty$

We prove in this chapter that a localization operator $L_{F,\varphi} : X \to X$ associated to a function F in $L^p(G)$, $1 \leq p \leq \infty$, and an admissible wavelet φ for an irreducible and square-integrable representation of a locally compact and Hausdorff group G on a Hilbert space X is in the Schatten-von Neumann class S_p, $1 \leq p \leq \infty$. When $p = 1$, the irreducibility of the representation $\pi : G \to U(X)$ can be dispensed with.

The first result on the Schatten-von Neumann property of localization operators is given in the following proposition.

Proposition 13.1 *Let $F \in L^1(G)$. Then the localization operator $L_{F,\varphi} : X \to X$ is in S_1 and*

$$\|L_{F,\varphi}\|_{S_1} \leq \frac{4}{c_\varphi}\|F\|_{L^1(G)}. \tag{13.1}$$

Proof. First we assume that the function $F : G \to \mathbb{C}$ is nonnegative and real-valued. Then, by (12.1), we get

$$(L_{F,\varphi}x, x) = \frac{1}{c_\varphi}\int_G F(g)|(x, \pi(g)\varphi)|^2 d\mu(g) \geq 0$$

for all x in X, *i.e.*, the localization operator $L_{F,\varphi} : X \to X$ is positive. Let $\{\varphi_k : k = 1, 2, \ldots\}$ be any orthonormal basis for X. Then, using (12.1) and the fact that $\pi : G \to U(X)$ is a representation, we get

$$\sum_{k=1}^{\infty}(L_{F,\varphi}\varphi_k, \varphi_k) = \sum_{k=1}^{\infty}\frac{1}{c_\varphi}\int_G F(g)|(\varphi_k, \pi(g)\varphi)|^2 d\mu(g). \tag{13.2}$$

Hence, using (13.2), Fubini's theorem, Parseval's identity, $\|\varphi\| = 1$ and the fact that $\pi(g) : X \to X$ is unitary for all g in G, we get

$$\sum_{k=1}^{\infty}(L_{F,\varphi}\varphi_k, \varphi_k) = \frac{1}{c_\varphi}\|F\|_{L^1(G)} < \infty. \tag{13.3}$$

Hence, by (13.3) and Proposition 2.4, the localization operator $L_{F,\varphi} : X \to X$ is in S_1. Furthermore,

$$(L_{F,\varphi}^* L_{F,\varphi})^{\frac{1}{2}} = L_{F,\varphi}. \tag{13.4}$$

Thus, if $\{\psi_k : k = 1, 2, \ldots\}$ is an orthonormal basis for X consisting of eigenvectors of $(L_{F,\varphi}^* L_{F,\varphi})^{\frac{1}{2}} : X \to X$, we have, by (13.2) and (13.4),

$$
\begin{aligned}
\|L_{F,\varphi}\|_{S_1} &= \sum_{k=1}^{\infty} ((L_{F,\varphi}^* L_{F,\varphi})^{\frac{1}{2}} \psi_k, \psi_k) \\
&= \sum_{k=1}^{\infty} (L_{F,\varphi} \psi_k, \psi_k) \\
&= \frac{1}{c_\varphi} \|F\|_{L^1(G)}.
\end{aligned}
\tag{13.5}
$$

Now, if $F \in L^1(G)$ is a real-valued function, then we write $F = F_+ - F_-$, where

$$
F_+(g) = \max(F(g), 0)
$$

and

$$
F_-(g) = -\min(F(g), 0)
$$

for all g in G. Then $L_{F,\varphi} : X \to X$ is in S_1 and by (13.5),

$$
\begin{aligned}
\|L_{F,\varphi}\|_{S_1} &= \|L_{F_+,\varphi} - L_{F_-,\varphi}\|_{S_1} \leq \|L_{F_+,\varphi}\|_{S_1} + \|L_{F_-,\varphi}\|_{S_1} \\
&= \frac{1}{c_\varphi} \left(\|F_+\|_{L^1(G)} + \|F_-\|_{L^1(G)} \right) \\
&\leq \frac{2}{c_\varphi} \|F\|_{L^1(G)}.
\end{aligned}
\tag{13.6}
$$

Finally, let $F \in L^1(G)$ be a complex-valued function. Then we write $F = F_1 + iF_2$, where F_1 and F_2 are the real and imaginary parts of F respectively. Then $L_{F,\varphi} : X \to X$ is a localization operator and by (13.6),

$$
\begin{aligned}
\|L_{F,\varphi}\|_{S_1} &= \|L_{F_1,\varphi} + iL_{F_2,\varphi}\|_{S_1} \leq \|L_{F_1,\varphi}\|_{S_1} + \|L_{F_2,\varphi}\|_{S_1} \\
&\leq \frac{2}{c_\varphi} \left(\|F_1\|_{L^1(G)} + \|F_2\|_{L^1(G)} \right) \\
&\leq \frac{4}{c_\varphi} \|F\|_{L^1(G)},
\end{aligned}
$$

and the proof of Proposition 13.1 is complete. $\square$

Remark 13.2 The very elementary proof of Proposition 13.1 gives us the constant $\frac{4}{c_\varphi}$ in the estimate (13.1). By Proposition 12.1, it is natural to expect that the best constant should be $\frac{1}{c_\varphi}$ instead of $\frac{4}{\varphi}$. That this is indeed the case is proved in Chapter 14 using more elaborate techniques.

A consequence of Proposition 13.1 is the following result.

Proposition 13.3 *Let $F \in L^p(G)$, $1 \leq p < \infty$. Then the localization operator $L_{F,\varphi} : X \to X$ is compact.*

Proof. We again denote by $\mathcal{D}$ the set of all simple functions F on G such that

$$\mu\{g \in G : F(g) \neq 0\} < \infty.$$

Let $\{F_k\}_{k=1}^{\infty}$ be a sequence of functions in $\mathcal{D}$ such that $F_k \to F$ in $L^p(G)$ as $k \to \infty$. Then, by (12.5),

$$\|L_{F_k,\varphi} - L_{F,\varphi}\|_* \leq \left(\frac{1}{c_\varphi}\right)^{\frac{1}{p}} \|F_k - F\|_{L^p(G)} \to 0$$

as $k \to \infty$, i.e., $L_{F_k,\varphi} \to L_{F,\varphi}$ in $B(X)$ as $k \to \infty$. Since, by Proposition 13.1, $L_{F_k,\varphi} : X \to X$ is in S_1 and hence compact, it follows that $L_{F,\varphi} : X \to X$ is compact. $\square$

Remark 13.4 That Proposition 13.3 is false for $p = \infty$ can be seen easily by taking the function F on G to be such that

$$F(g) = 1, \quad g \in G.$$

For then, by the resolution of the identity formula in Theorem 6.1, $L_{F,\varphi} : X \to X$ is the identity operator on X. In view of the hypothesis that X is infinite-dimensional, $L_{F,\varphi} : X \to X$ is not compact.

Now, we can come to the main result on the Schatten-von Neumann property of localization operators.

Theorem 13.5 *Let $F \in L^p(G)$, $1 \leq p \leq \infty$. Then the localization operator $L_{F,\varphi} : X \to X$ is in S_p and*

$$\|L_{F,\varphi}\|_{S_p} \leq \left(\frac{4}{c_\varphi}\right)^{\frac{1}{p}} \|F\|_{L^p(G)}.$$

Proof. By Proposition 13.1,

$$\|L_{F,\varphi}\|_{S_1} \leq \frac{4}{c_\varphi}\|F\|_{L^1(G)}, \quad F \in L^1(G), \tag{13.7}$$

and, by Proposition 12.2,

$$\|L_{F,\varphi}\|_{S_\infty} = \|L_{F,\varphi}\|_* \leq \|F\|_{L^\infty(G)}, \quad F \in L^\infty(G). \tag{13.8}$$

So, by (13.7), (13.8), Theorems 2.10 and 2.11, the proof is complete. $\square$

We end this chapter with a formula for the trace $\text{tr}(L_{F,\varphi})$ of the localization operator $L_{F,\varphi} : X \to X$ when $F \in L^1(G)$.

Theorem 13.6 *Let $F \in L^1(G)$. Then* $\operatorname{tr}(L_{F,\varphi}) = \frac{1}{c_\varphi} \int_G F(g) d\mu(g)$.

Proof. Let $\{\varphi_k : k = 1, 2, \ldots\}$ be an orthonormal basis for X. Then, using the definition of the trace given in Chapter 2, the definition of $L_{F,\varphi} : X \to X$, Fubini's theorem, Parseval's identity, $\|\varphi\| = 1$ and the fact that $\pi(g) : X \to X$ is a unitary operator for all g in G, we get

$$
\begin{aligned}
\operatorname{tr}(L_{F,\varphi}) &= \sum_{k=1}^{\infty} (L_{F,\varphi} \varphi_k, \varphi_k) \\
&= \sum_{k=1}^{\infty} \frac{1}{c_\varphi} \int_G F(g) |(\varphi_k, \pi(g)\varphi)|^2 d\mu(g) \\
&= \frac{1}{c_\varphi} \int_G F(g) \sum_{k=1}^{\infty} |(\varphi_k, \pi(g)\varphi)|^2 d\mu(g) \\
&= \frac{1}{c_\varphi} \int_G F(g) \|\pi(g)\varphi\|^2 d\mu(g) \\
&= \frac{1}{c_\varphi} \int_G F(g) d\mu(g).
\end{aligned}
$$

$\square$

14 Trace Class Norm Inequalities

As a sequel to Chapter 13, we prove in this chapter that the constant $\frac{4}{c_\varphi}$ in Proposition 13.1 can be improved to $\frac{1}{c_\varphi}$ and obtain a lower bound for the norm $\|L_{F,\varphi}\|_{S_1}$ of the localization operator $L_{F,\varphi} : X \to X$ in S_1 in terms of the function F_φ on G defined by

$$F_\varphi(g) = (L_{F,\varphi}\pi(g)\varphi, \pi(g)\varphi), \quad g \in G. \tag{14.1}$$

The function F_φ can be interpreted as the expectation value of the "observable" $L_{F,\varphi} : X \to X$ in the coherent states $\pi(g)\varphi$, $g \in G$. See, for instance, the book [2] by Ali, Antoine and Gazeau, the survey paper [3] by Ali, Antoine, Gazeau and Mueller, and the book [4] by Berezin and Shubin for comprehensive accounts of coherent states and related topics.

An important result in this chapter is the following theorem on the norm $\|L_{F,\varphi}\|_{S_1}$ of the localization operator $L_{F,\varphi} : X \to X$ in S_1 when $F \in L^1(G)$.

Theorem 14.1 *Let $F \in L^1(G)$. Then*

$$\|L_{F,\varphi}\|_{S_1} \le \frac{1}{c_\varphi}\|F\|_{L^1(G)}.$$

If, in addition, the square-integrable representation $\pi : G \to U(X)$ of G on X is irreducible, then

$$\frac{1}{c_\varphi}\|F_\varphi\|_{L^1(G)} \le \|L_{F,\varphi}\|_{S_1}.$$

Remark 14.2 To see that $F_\varphi \in L^1(G)$, we use (12.1), (14.1), Fubini's theorem, $\|\varphi\| = 1$, the fact that $\pi : G \to U(X)$ is a unitary representation, the left invariance of μ and the definition of the wavelet constant c_φ to get

$$
\begin{aligned}
\int_G |F_\varphi(g)|d\mu(g) &= \int_G \left| \frac{1}{c_\varphi} \int_G F(h)|(\pi(g)\varphi, \pi(h)\varphi)|^2 d\mu(h) \right| d\mu(g) \\
&\le \frac{1}{c_\varphi} \int_G \left(\int_G |F(h)||(\pi(g)\varphi, \pi(h)\varphi)|^2 d\mu(g) \right) d\mu(h) \\
&= \frac{1}{c_\varphi} \int_G |F(h)| \left(\int_G |(\pi(g)\varphi, \pi(h)\varphi)|^2 d\mu(g) \right) d\mu(h) \\
&= \frac{1}{c_\varphi} \int_G |F(h)| \left(\int_G |(\pi(h^{-1}g)\varphi, \varphi)|^2 d\mu(g) \right) d\mu(h) \\
&= \int_G |F(h)|d\mu(h) < \infty.
\end{aligned}
$$

The following connection between F and F_φ is also interesting in its own right.

Proposition 14.3 *Let $F \in L^1(G)$. Then*

$$\int_G F_\varphi(g)d\mu(g) = \int_G F(g)d\mu(g).$$

Proof. By (12.1), (14.1), Fubini's theorem, $\|\varphi\| = 1$, the fact that $\pi : G \to U(X)$ is a unitary representation, the left invariance of μ and the definition of the wavelet constant c_φ, we get

$$
\begin{aligned}
\int_G F_\varphi(g)d\mu(g) &= \frac{1}{c_\varphi} \int_G \left(\int_G F(h)|(\pi(g)\varphi, \pi(h)\varphi)|^2 d\mu(h) \right) d\mu(g) \\
&= \frac{1}{c_\varphi} \int_G F(h) \left(\int_G |(\pi(g)\varphi, \pi(h)\varphi)|^2 d\mu(g) \right) d\mu(h) \\
&= \frac{1}{c_\varphi} \int_G F(h) \left(\int_G |(\pi(h^{-1}g)\varphi, \varphi)|^2 d\mu(g) \right) d\mu(h) \\
&= \int_G F(h)d\mu(h).
\end{aligned}
$$

$\square$

Remark 14.4 By Theorem 13.6, we know that

$$\text{tr}(L_{F,\varphi}) = \frac{1}{c_\varphi} \int_G F(g)d\mu(g).$$

If F is a real-valued and nonnegative function in $L^1(G)$, then $L_{F,\varphi} : X \to X$ is a positive operator and hence $\|L_{F,\varphi}\|_{S_1} = \text{tr}(L_{F,\varphi})$. Hence, by Proposition 14.3,

$$\frac{1}{c_\varphi} \int_G F_\varphi(g)d\mu(g) = \|L_{F,\varphi}\|_{S_1} = \frac{1}{c_\varphi} \int_G F(g)d\mu(g).$$

Thus, the estimates in Theorem 14.1 are sharp.

Proof of Theorem 14.1. We note that, by Proposition 13.1, $L_{F,\varphi} \in S_1$. Then, using the canonical form for compact operators given in Theorem 2.2, we get

$$L_{F,\varphi}x = \sum_{k=1}^\infty s_k(L_{F,\varphi})(x, \varphi_k)\psi_k, \quad x \in X, \tag{14.2}$$

where $s_k(L_{F,\varphi})$, $k = 1, 2, \ldots$, are the positive singular values of $L_{F,\varphi} : X \to X$, $\{\varphi_k : k = 1, 2, \ldots\}$ is an orthonormal basis for $N(L_{F,\varphi})^\perp$, $\{\psi_k : k = 1, 2, \ldots\}$ is an orthonormal set in X and the convergence of the series is understood to be in X. By (14.2),

$$\sum_{j=1}^{\infty}(L_{F,\varphi}\varphi_j,\psi_j) = \sum_{j=1}^{\infty}s_j(L_{F,\varphi}). \tag{14.3}$$

So, using the definition of $\|\ \|_{S_1}$ and (14.3),

$$\|L_{F,\varphi}\|_{S_1} = \sum_{j=1}^{\infty}(L_{F,\varphi}\varphi_j,\psi_j). \tag{14.4}$$

Thus, by (12.1), (14.4), Fubini's theorem, Schwarz' inequality, Bessel's inequality, $\|\varphi\| = 1$ and the fact that $\pi : G \to U(X)$ is a unitary representation, we get

$$
\begin{aligned}
\|L_{F,\varphi}\|_{S_1} &= \sum_{k=1}^{\infty}(L_{F,\varphi}\varphi_k,\psi_k) \\
&= \sum_{k=1}^{\infty}\frac{1}{c_\varphi}\int_G F(g)(\varphi_k,\pi(g)\varphi)(\pi(g)\varphi,\psi_k)d\mu(g) \\
&\le \sum_{k=1}^{\infty}\frac{1}{c_\varphi}\int_G |F(g)||(\varphi_k,\pi(g)\varphi)||(\pi(g)\varphi,\psi_k)|d\mu(g) \\
&= \frac{1}{c_\varphi}\int_G |F(g)|\sum_{k=1}^{\infty}|(\varphi_k,\pi(g)\varphi)||(\pi(g)\varphi,\psi_k)|d\mu(g) \\
&\le \frac{1}{c_\varphi}\int_G |F(g)|\left\{\sum_{j=1}^{\infty}|(\varphi_j,\pi(g)\varphi)|^2\sum_{k=1}^{\infty}|(\pi(g)\varphi,\psi_k)|^2\right\}^{\frac{1}{2}}d\mu(g) \\
&\le \frac{1}{c_\varphi}\int_G |F(g)|\,\|\pi(g)\varphi\|^2 d\mu(g) \\
&= \frac{1}{c_\varphi}\int_G |F(g)|d\mu(g) = \frac{1}{c_\varphi}\|F\|_{L^1(G)}.
\end{aligned}
$$

Also, by (14.1) and (14.2),

$$
\begin{aligned}
|F_\varphi(g)| &= |(L_{F,\varphi}\pi(g)\varphi,\pi(g)\varphi)| \\
&= \left|\sum_{k=1}^{\infty}s_k(L_{F,\varphi})(\pi(g)\varphi,\varphi_k)(\psi_k,\pi(g)\varphi)\right| \\
&\le \frac{1}{2}\sum_{k=1}^{\infty}s_k(L_{F,\varphi})(|(\pi(g)\varphi,\varphi_k)|^2 + |(\psi_k,\pi(g)\varphi)|^2) \tag{14.5}
\end{aligned}
$$

for all g in G. So, using (14.5), Fubini's theorem, the irreducibility of the representation $\pi : G \to U(X)$ and hence the resolution of the identity formula in

Theorem 6.1, and

$$\|\varphi_k\| = \|\psi_k\| = 1, \quad k = 1, 2, \ldots,$$

we get

$$\int_G |F_\varphi(g)| d\mu(g)$$

$$\leq \frac{1}{2} \sum_{k=1}^{\infty} s_k(L_{F,\varphi}) \left\{ \int_G |(\pi(g)\varphi, \varphi_k)|^2 d\mu(g) + \int_G |(\psi_k, \pi(g)\varphi)|^2 d\mu(g) \right\}$$

$$= c_\varphi \sum_{k=1}^{\infty} s_k(L_{F,\varphi}). \tag{14.6}$$

Thus, using the definition of $\| \ \|_{S_1}$ and (14.6),

$$\frac{1}{c_\varphi} \int_G |F_\varphi(g)| d\mu(g) \leq \|L_{F,\varphi}\|_{S_1},$$

and the proof is complete.   □

An immediate consequence of Theorem 14.1 is the following improvement of Theorem 13.5.

Theorem 14.5 *Let $F \in L^p(G)$, $1 \leq p \leq \infty$. Then the localization operator $L_{F,\varphi} : X \to X$ is in S_p and*

$$\|L_{F,\varphi}\|_{S_p} \leq \left(\frac{1}{c_\varphi} \right)^{\frac{1}{p}} \|F\|_{L^p(G)}.$$

The proof of Theorem 14.5 is the same as the proof of Theorem 13.5 if we replace the S_1-estimate by

$$\|L_{F,\varphi}\|_{S_1} \leq \frac{1}{c_\varphi} \|F\|_{L^1(G)}.$$

We are now in a position to derive optimal upper and lower bounds for the norms in S_1 of convex functions of self-adjoint localization operators. We begin with a brief recall of convex functions and Jensen's inequality.

A real-valued function v defined on an interval (a, b), where $-\infty \leq a < b \leq \infty$, is said to be convex if

$$v((1 - \lambda)x + \lambda y) \leq (1 - \lambda)v(x) + \lambda v(y)$$

for all x and y in (a, b) and all λ in $[0, 1]$. A fundamental property of convex functions is provided by the following theorem.

Theorem 14.6 Let $f \in L^1(\Omega, \mu)$, where (Ω, μ) is a probability measure space, i.e., $\mu(\Omega) = 1$. Let f be a real-valued function in $L^1(\Omega, \mu)$ such that the range of f is contained in an open interval (a, b), where $-\infty \le a < b \le \infty$. If v is a convex function on (a, b), then

$$v\left(\int_\Omega f(\omega)d\mu(\omega)\right) \le \int_\Omega (v \circ f)(\omega)d\mu(\omega).$$

Remark 14.7 We allow $a = -\infty$ or $b = \infty$. The function $v \circ f$ may fail to be in $L^1(\Omega, \mu)$, and in this case,

$$\int_\Omega (v \circ f)(\omega)d\mu(\omega) = \infty.$$

The inequality in Theorem 14.6 is known as Jensen's inequality. A proof can be found on pages 62 and 63 of the book [74] by Rudin.

Let F be a real-valued function in $L^1(G)$. Then, by Proposition 13.1, the localization operator $L_{F,\varphi} : X \to X$ is in S_1. Furthermore, it is self-adjoint. Indeed, for all x and y in X, we can use (12.1) and get

$$
\begin{aligned}
(L_{F,\varphi}x, y) &= \frac{1}{c_\varphi} \int_G F(g)(x, \pi(g)\varphi)(\pi(g)\varphi, y)d\mu(g) \\
&= \overline{\frac{1}{c_\varphi} \int_G F(g)(y, \pi(g)\varphi)(\pi(g)\varphi, x)d\mu(g)} \\
&= \overline{(L_{F,\varphi}y, x)} = (x, L_{F,\varphi}y).
\end{aligned}
$$

Hence, using the spectral theorem for compact and self-adjoint operators, we get

$$L_{F,\varphi}x = \sum_{j=1}^\infty \lambda_j(x, \varphi_j)\varphi_j, \quad x \in X, \tag{14.7}$$

where $\{\varphi_j : j = 1, 2, \ldots\}$ is an orthonormal basis for X consisting of eigenvectors of $L_{F,\varphi} : X \to X$ and λ_j is the eigenvalue of $L_{F,\varphi} : X \to X$ corresponding to the eigenvector φ_j, $j = 1, 2, \ldots$, and the convergence is understood to be in X.

Let $H : \mathbb{R} \to \mathbb{R}$ be a convex function. Then H is a bounded function on compact subsets of $\mathbb{R}$. We define the bounded linear operator $H(L_{F,\varphi}) : X \to X$ by

$$(H(L_{F,\varphi})x, y) = \sum_{j=1}^\infty H(\lambda_j)(x, \varphi_j)(\varphi_j, y) \tag{14.8}$$

for all x and y in X. That $H(L_{F,\varphi}) : X \to X$ is a bounded linear operator is easy to check. Indeed, we use the fact that H is a bounded function on compact subsets of $\mathbb{R}$, Schwarz' inequality and Parseval's identity to get

$$|(H(L_{F,\varphi})x, y| \;\leq\; \sum_{j=1}^{\infty} |H(\lambda_j)||(x, \varphi_j)||(\varphi_j, y)|$$

$$\leq \;\; \sup_{\lambda \in \Sigma(L_{F,\varphi})} |H(\lambda)| \left\{ \sum_{j=1}^{\infty} |(x, \varphi_j)|^2 \right\}^{\frac{1}{2}} \left\{ \sum_{j=1}^{\infty} |(y, \varphi_j)|^2 \right\}^{\frac{1}{2}}$$

$$= \;\; \sup_{\lambda \in \Sigma(L_{F,\varphi})} |H(\lambda)| \, \|x\| \, \|y\|,$$

where $\Sigma(L_{F,\varphi})$ is the spectrum of $L_{F,\varphi} : X \to X$.

Proposition 14.8 $H(L_{F,\varphi}) : X \to X$ *is a self-adjoint operator.*

Proof. For all x and y in X, we get

$$(H(L_{F,\varphi})x, y) \;=\; \sum_{j=1}^{\infty} H(\lambda_j)(x, \varphi_j)(\varphi_j, y)$$

$$= \;\; \sum_{j=1}^{\infty} \overline{H(\lambda_j)(y, \varphi_j)(\varphi_j, x)}$$

$$= \;\; \overline{(H(L_{F,\varphi})y, x)} = (x, H(L_{F,\varphi})y).$$

$\square$

Proposition 14.9 *For* $j = 1, 2, \ldots$, $H(\lambda_j)$ *is the eigenvalue of* $H(L_{F,\varphi}) : X \to X$ *corresponding to the eigenvector* φ_j.

Proof. Since $\{\varphi_k : k = 1, 2, \ldots\}$ is orthonormal, it follows that

$$(H(L_{F,\varphi})\varphi_j, x) = \sum_{k=1}^{\infty} H(\lambda_k)(\varphi_j, \varphi_k)(\varphi_k, x) = (H(\lambda_j)\varphi_j, x)$$

for $j = 1, 2, \ldots$, and all x in X.

$\square$

We can now give a necessary and sufficient condition for the bounded linear operator $H(L_{F,\varphi}) : X \to X$ to be in S_1 in terms of the convex function $H : \mathbb{R} \to \mathbb{R}$. In the case when $H(L_{F,\varphi}) : X \to X$ is in S_1, we can give an upper bound and a lower bound for the norm $\|H(L_{F,\varphi})\|_{S_1}$ of $H(L_{F,\varphi}) : X \to X$ in S_1 in terms of the functions $H \circ F : G \to \mathbb{R}$ and $H \circ F_\varphi : G \to \mathbb{R}$ respectively, where the function F_φ is defined by (14.1).

Theorem 14.10 *The bounded linear operator* $H(L_{F,\varphi}) : X \to X$ *is in* S_1 *if and only if*

$$\sum_{j=1}^{\infty} |H(\lambda_j)| < \infty.$$

Proof. By Proposition 14.9 and the spectral mapping theorem, $|H(\lambda_j)|$, $j = 1, 2, \ldots$, are the singular values of $H(L_{F,\varphi}) : X \to X$. Thus, using the definition of S_1, the proof is complete. $\qquad\Box$

Now, we define $|H(L_{F,\varphi})|_{S_1}$ by

$$|H(L_{F,\varphi})|_{S_1} = \begin{cases} \|H(L_{F,\varphi})\|_{S_1} & \text{if } \sum_{j=1}^{\infty} |H(\lambda_j)| < \infty, \\ \infty & \text{if } \sum_{j=1}^{\infty} |H(\lambda_j)| = \infty, \end{cases} \tag{14.9}$$

and we have the following theorem.

Theorem 14.11 *Let $F \in L^1(G)$ be a real-valued function, and let $H : \mathbb{R} \to \mathbb{R}$ be a nonnegative and convex function. Then*

$$\frac{1}{c_\varphi} \|H \circ F_\varphi\|_{L^1(G)} \leq |H(L_{F,\varphi})|_{S_1} \leq \frac{1}{c_\varphi} \|H \circ F\|_{L^1(G)},$$

where $F_\varphi : G \to \mathbb{R}$ is defined by (14.1).

Remark 14.12 If $H(\lambda) = |\lambda|$ for all $\lambda \in \mathbb{R}$, then Theorem 14.11 is a special case of Theorem 14.1. Thus, by Remark 14.4, Theorem 14.11 is sharp.

Remark 14.13 The inequalities in Theorem 14.11 are refined analogues of the inequalities (2.74) in Chapter 5 of the book [4] by Berezin and Shubin for localization operators. The origin of these inequalities can be traced back to the study of a Feynman inequality in quantization. See the paper [91] by Symanzik for details.

Proof of Theorem 14.11. Let $x \in X$ be such that $\|x\| = 1$. Then, by (14.7), (14.8) and Jensen's inequality,

$$\begin{aligned} (H(L_{F,\varphi})x, x) &= \sum_{j=1}^{\infty} H(\lambda_j)|(x, \varphi_j)|^2 \\ &\geq H\left(\sum_{j=1}^{\infty} \lambda_j |(x, \varphi_j)|^2\right) = H((L_{F,\varphi}x, x)). \end{aligned} \tag{14.10}$$

In (14.10), we let $x = \pi(g)\varphi$, $g \in G$, integrate on G and use (14.1). Then we get

$$\begin{aligned} \int_G (H(L_{F,\varphi})\pi(g)\varphi, \pi(g)\varphi)d\mu(g) &\geq \int_G H((L_{F,\varphi}\pi(g)\varphi, \pi(g)\varphi))d\mu(g) \\ &= \int_G H(F_\varphi(g))d\mu(g). \end{aligned} \tag{14.11}$$

Now, by (14.8), (14.9), Fubini's theorem, the resolution of the identity formula in Theorem 6.1 and the fact that

$$\|\varphi_j\| = 1, \quad j = 1, 2, \ldots,$$

we get

$$\int_G (H(L_{F,\varphi})\pi(g)\varphi, \pi(g)\varphi)d\mu(g)$$

$$\leq \int_G \sum_{j=1}^{\infty} |H(\lambda_j)||(\pi(g)\varphi, \varphi_j)|^2 d\mu(g)$$

$$= \sum_{j=1}^{\infty} |H(\lambda_j)| \int_G |(\pi(g)\varphi, \varphi_j)|^2 d\mu(g)$$

$$= \sum_{j=1}^{\infty} |H(\lambda_j)| c_\varphi \|\varphi_j\|^2 = c_\varphi \sum_{j=1}^{\infty} |H(\lambda_j)|$$

$$= c_\varphi |H(L_{F,\varphi})|_{S_1}. \tag{14.12}$$

So, by (14.11) and (14.12),

$$\frac{1}{c_\varphi}\|H \circ F_\varphi\|_{L^1(G)} \leq |H(L_{F,\varphi})|_{S_1}. \tag{14.13}$$

Next, by (12.1) and (14.9),

$$|H(L_{F,\varphi})|_{S_1} = \sum_{j=1}^{\infty} |H(\lambda_j)| = \sum_{j=1}^{\infty} |H(L_{F,\varphi}\varphi_j, \varphi_j)|$$

$$= \sum_{j=1}^{\infty} \left| H\left(\frac{1}{c_\varphi}\int_G F(g)|(\varphi_j, \pi(g)\varphi)|^2 d\mu(g)\right)\right|. \tag{14.14}$$

Since

$$\|\varphi_j\| = 1, \quad j = 1, 2, \ldots,$$

it follows from the resolution of the identity formula in Theorem 6.1 again that

$$\frac{1}{c_\varphi}\int_G |(\varphi_j, \pi(g)\varphi)|^2 d\mu(g) = \|\varphi_j\|^2 = 1 \tag{14.15}$$

for $j = 1, 2, \ldots$. Thus, by (14.14), (14.15), Jensen's inequality and Fubini's theorem,

$$|H(L_{F,\varphi})|_{S_1} \leq \frac{1}{c_\varphi}\sum_{j=1}^{\infty}\int_G |H(F(g))| |(\varphi_j, \pi(g)\varphi)|^2 d\mu(g)$$

$$= \frac{1}{c_\varphi}\int_G |H(F(g))| \sum_{j=1}^{\infty} |(\varphi_j, \pi(g)\varphi)|^2 d\mu(g). \tag{14.16}$$

So, by (14.16), Parseval's identity, the fact that $\pi : G \to U(X)$ is a unitary representation and $\|\varphi\| = 1$, we get

$$|H(L_{F,\varphi})|_{S_1} \leq \frac{1}{c_\varphi}\int_G |H(F(g))| \|\pi(g)\varphi\|^2 d\mu(g) = \frac{1}{c_\varphi}\int_G |H(F(g))|d\mu(g). \tag{14.17}$$

Thus, by (14.13) and (14.17), the proof is complete.   □

15 Hilbert-Schmidt Localization Operators

Let $\pi : G \to U(X)$ be an irreducible and square-integrable representation of a locally compact and Hausdorff group G on a Hilbert space X. Then for all functions F in $L^p(G)$, $1 \leq p \leq \infty$, and all admissible wavelets φ for $\pi : G \to U(X)$, Proposition 12.3 ensures that we can get a unique bounded linear operator $L_{F,\varphi} : X \to X$ such that

$$\|L_{F,\varphi}\|_* \leq \left(\frac{1}{c_\varphi}\right)^{\frac{1}{p}} \|F\|_{L^p(G)} \tag{15.1}$$

and

$$(L_{F,\varphi}x, y) = \frac{1}{c_\varphi} \int_G F(g)(x, \pi(g)\varphi)(\pi(g)\varphi, y)d\mu(g) \tag{15.2}$$

for all x and y in X whenever F is a simple function on G for which

$$\mu\{g \in G : F(g) \neq 0\} < \infty.$$

Furthermore, Theorem 14.5 tells us that $L_{F,\varphi} : X \to X$ is in the Schatten-von Neumann class S_p and

$$\|L_{F,\varphi}\|_{S_p} \leq \left(\frac{1}{c_\varphi}\right)^{\frac{1}{p}} \|F\|_{L^p(G)}. \tag{15.3}$$

These results entail the use of the Riesz-Thorin theorem and Theorems 2.10 and 2.11 in the theory of interpolation. However, when $p = 2$, we can give an explicit formula, i.e., (15.2), for the localization operator $L_{F,\varphi} : X \to X$ for all F in $L^2(G)$ and prove that it is a bounded linear operator. If, in addition, the group G is unimodular, then we can prove that $L_{F,\varphi} : X \to X$ is in the Hilbert-Schmidt class S_2 satisfying (15.1) and (15.3) without using interpolation theory.

Let us begin with the following proposition.

Proposition 15.1 *The localization operator $L_{F,\varphi} : X \to X$ defined by (15.2) for all x and y in X and all F in $L^2(G)$ is a bounded linear operator and*

$$\|L_{F,\varphi}\|_* \leq \left(\frac{1}{c_\varphi}\right)^{\frac{1}{2}} \|F\|_{L^2(G)}.$$

Proof. For all x and y in X, the function $H : G \to \mathbb{C}$ defined by

$$H(g) = (x, \pi(g)\varphi)(\pi(g)\varphi, y), \quad g \in G, \tag{15.4}$$

is in $L^2(G)$. Indeed, using Schwarz' inequality, $\|\varphi\| = 1$ and the fact that $\pi : G \to U(X)$ is a unitary representation of G on X, we get

$$|(\pi(g)\varphi, y)| \le \|y\|, \quad g \in G. \tag{15.5}$$

So, using (15.5) and the resolution of identity formula in Theorem 6.1,

$$
\begin{aligned}
\int_G |H(g)|^2 d\mu(g) &= \int_G |(x, \pi(g)\varphi)(\pi(g)\varphi, y)|^2 d\mu(g) \\
&\le \left(\int_G |(x, \pi(g)\varphi)|^2 d\mu(g) \right) \|y\|^2 \\
&= c_\varphi \|x\|^2 \|y\|^2.
\end{aligned}
\tag{15.6}
$$

Thus, by (15.2), (15.4), (15.6), Schwarz' inequality and the resolution of the identity formula in Theorem 6.1,

$$|(L_{F,\varphi}x, y)| \le \frac{1}{c_\varphi} \|F\|_{L^2(G)} \|H\|_{L^2(G)} \le \left(\frac{1}{c_\varphi} \right)^{\frac{1}{2}} \|F\|_{L^2(G)} \|x\| \|y\|,$$

and the proof is complete. □

Proposition 15.2 *Let G be a unimodular group. Then for all F in $L^2(G)$, the localization operator $L_{F,\varphi} : X \to X$ is in the Hilbert-Schmidt class S_2 and*

$$\|L_{F,\varphi}\|_{S_2} \le \left(\frac{1}{c_\varphi} \right)^{\frac{1}{2}} \|F\|_{L^2(G)}.$$

Proposition 15.2 is definitely much weaker than Theorem 14.5 in which the unimodularity of the group G is not needed. However, it is instructive to see a proof of the Hilbert-Schmidt property of localization operators from first principles without using interpolation theory. The price for this elementary proof is unimodularity.

To prove Proposition 15.2, we use the following lemma, which follows from (15.2) immediately.

Lemma 15.3 *The adjoint of $L_{F,\varphi} : X \to X$ is $L_{\overline{F},\varphi} : X \to X$.*

Proof of Proposition 15.2. Let $F \in L^1(G) \cap L^2(G)$. Let $\{\varphi_k : k = 1, 2, \ldots\}$ be an orthonormal basis for X. Then, by (15.2), Lemma 15.3, Fubini's theorem and

Parseval's identity,

$$
\sum_{k=1}^{\infty}\|L_{F,\varphi}\varphi_k\|^2 = \sum_{k=1}^{\infty}(L_{F,\varphi}\varphi_k, L_{F,\varphi}\varphi_k)
$$

$$
= \sum_{k=1}^{\infty}\frac{1}{c_\varphi}\int_G F(g)(\varphi_k, \pi(g)\varphi)(\pi(g)\varphi, L_{F,\varphi}\varphi_k)d\mu(g)
$$

$$
= \sum_{k=1}^{\infty}\frac{1}{c_\varphi}\int_G F(g)(\varphi_k, \pi(g)\varphi)(L_{\overline{F},\varphi}\pi(g)\varphi, \varphi_k)d\mu(g)
$$

$$
= \frac{1}{c_\varphi}\int_G F(g)(L_{\overline{F},\varphi}\pi(g)\varphi, \pi(g)\varphi)d\mu(g). \tag{15.7}
$$

Hence, by (15.5), (15.7), Fubini's theorem, Schwarz' inequality, the unimodularity of G, the fact that $\pi : G \to U(X)$ is an irreducible and square-integrable representation, and $\|x\| = 1$, we get

$$
\sum_{k=1}^{\infty}\|L_{F,\varphi}\varphi_k\|^2
$$

$$
= \left(\frac{1}{c_\varphi}\right)^2\int_G F(g)\left(\int_G \overline{F}(h)|(\pi(h)\varphi, \pi(g)\varphi)|^2 d\mu(h)\right)d\mu(g)
$$

$$
= \left(\frac{1}{c_\varphi}\right)^2\int_G F(g)\left(\int_G \overline{F}(h)|(\pi(g^{-1}h)\varphi, \varphi)|^2 d\mu(h)\right)d\mu(g)
$$

$$
= \left(\frac{1}{c_\varphi}\right)^2\int_G F(g)\left(\int_G \overline{F}(gh)|(\pi(h)\varphi, \varphi)|^2 d\mu(h)\right)d\mu(g)
$$

$$
= \left(\frac{1}{c_\varphi}\right)^2\int_G |(\pi(h)\varphi, \varphi)|^2\left(\int_G F(g)\overline{F}(gh)d\mu(g)\right)d\mu(h)
$$

$$
\leq \frac{1}{c_\varphi}\|F\|_{L^2(G)}^2. \tag{15.8}
$$

Thus, by (15.8), we get

$$
\|L_{F,\varphi}\|_{S_2} \leq \left(\frac{1}{c_\varphi}\right)^{\frac{1}{2}}\|F\|_{L^2(G)} \tag{15.9}
$$

for all F in $L^1(G) \cap L^2(G)$. Now, let $F \in L^2(G)$. Then there exists a sequence $\{F_l\}_{l=1}^{\infty}$ of functions in $L^1(G) \cap L^2(G)$ such that

$$
F_l \to F \tag{15.10}
$$

in $L^2(G)$ as $l \to \infty$. By (15.10) and Proposition 15.1,

$$
L_{F_l,\varphi} \to L_{F,\varphi} \tag{15.11}
$$

in $B(X)$ as $l \to \infty$. By (15.9) and (15.10),

$$\|L_{F_l,\varphi} - A\|_{S_2} \to 0 \tag{15.12}$$

for some A in S_2 as $l \to \infty$. Now, we claim that $L_{F_l,\varphi} \to A$ in $B(X)$ as $l \to \infty$. Assume that this is true for a moment. Then, by (15.11),

$$L_{F,\varphi} = A. \tag{15.13}$$

Hence $L_{F,\varphi} : X \to X$ is in S_2 and, by (15.9), (15.10), (15.12) and (15.13),

$$\|L_{F,\varphi}\|_{S_2} = \lim_{l \to \infty} \|L_{F_l,\varphi}\|_{S_2} \leq \left(\frac{1}{c_\varphi}\right)^{\frac{1}{2}} \lim_{l \to \infty} \|F_l\|_{L^2(G)} = \left(\frac{1}{c_\varphi}\right)^{\frac{1}{2}} \|F\|_{L^2(G)}.$$

It remains to prove that $L_{F_l,\varphi} \to A$ in $B(X)$ as $l \to \infty$. By Schwarz' inequality, Fubini's theorem and Parseval's identity,

$$\|(L_{F_l,\varphi} - A)x\|^2$$

$$= \left((L_{F_l,\varphi} - A) \sum_{j=1}^\infty (x, \varphi_j)\varphi_j, (L_{F_l,\varphi} - A) \sum_{k=1}^\infty (x, \varphi_k)\varphi_k \right)$$

$$\leq \sum_{j=1}^\infty \sum_{k=1}^\infty |(x, \varphi_j)| \, |(x, \varphi_k)| \, \|(L_{F_l,\varphi} - A)\varphi_j\| \, \|(L_{F_l,\varphi} - A)\varphi_k\|$$

$$= \left(\sum_{j=1}^\infty |(x, \varphi_j)| \, \|(L_{F_l,\varphi} - A)\varphi_j\| \right)^2$$

$$\leq \left(\sum_{j=1}^\infty |(x, \varphi_j)|^2 \right) \left(\sum_{j=1}^\infty \|(L_{F_l,\varphi} - A)\varphi_j\|^2 \right)$$

$$= \|L_{F_l,\varphi} - A\|_{S_2}^2 \|x\|^2$$

for all x in X, and hence

$$\|L_{F_l,\varphi} - A\|_* \leq \|L_{F_l,\varphi} - A\|_{S_2} \tag{15.14}$$

for $l = 1, 2, \ldots$. Thus, by (15.12), (15.13) and (15.14), $L_{F_l,\varphi} \to A$ in $B(X)$ as $l \to \infty$. $\square$

We end this chapter with a formula for the norm $\|L_{F,\varphi}\|_{S_2}$ of the Hilbert-Schmidt localization operator $L_{F,\varphi} : X \to X$. To this end, we let F^φ be the function on G defined by

$$F^\varphi(g) = \|L_{F,\varphi}\pi(g)\varphi\|, \quad g \in G. \tag{15.15}$$

In fact, for all g in G,

$$(F^\varphi(g))^2 = (L^*_{F,\varphi} L_{F,\varphi} \pi(g)\varphi, \pi(g)\varphi).$$

Thus, the square of the function F^φ can be considered as the expectation value of the observable $L^*_{F,\varphi} L_{F,\varphi} : X \to X$ in the coherent states $\pi(g)\varphi$, $g \in G$.

Theorem 15.4 *Let $F \in L^2(G)$. Then*

$$\|L_{F,\varphi}\|_{S_2} = \left(\frac{1}{c_\varphi}\right)^{\frac{1}{2}} \|F^\varphi\|_{L^2(G)}.$$

Proof. By Theorem 14.5, $L_{F,\varphi} \in S_2$. Using the canonical form for compact operators given by Theorem 2.2, we can write

$$L_{F,\varphi} x = \sum_{k=1}^{\infty} s_k(L_{F,\varphi})(x, \varphi_k)\psi_k, \quad x \in X, \tag{15.16}$$

where $s_k(L_{F,\varphi})$, $k = 1, 2, \ldots$, are the positive singular values of $L_{F,\varphi} : X \to X$, $\{\varphi_k : k = 1, 2, \ldots\}$ is an orthonormal basis for $N(L_{F,\varphi})^\perp$, and $\{\psi_k : k = 1, 2, \ldots\}$ is an orthonormal set in X. Thus, by (15.15) and (15.16), we get

$$(F^\varphi(g))^2 = \sum_{k=1}^{\infty} (s_k(L_{F,\varphi}))^2 |(\pi(g)\varphi, \varphi_k)|^2, \quad g \in G. \tag{15.17}$$

So, by (15.17),

$$\int_G (F^\varphi(g))^2 d\mu(g) = \sum_{k=1}^{\infty} (s_k(L_{F,\varphi}))^2 \int_G |(\pi(g)\varphi, \varphi_k)|^2 d\mu(g). \tag{15.18}$$

Hence, applying the resolution of the identity formula in Theorem 6.1 to the right-hand side of (15.18), we complete the proof. $\qquad\square$

16 Two-Wavelet Theory

The results hitherto given are for localization operators $L_{F,\varphi} : X \to X$ defined in terms of one admissible wavelet φ for the square-integrable representation $\pi : G \to U(X)$ of G on X. In this chapter we introduce the notion of a localization operator $L_{F,\varphi,\psi} : X \to X$, which is defined in terms of a symbol F in $L^1(G)$ and two admissible wavelets φ and ψ for the square-integrable representation $\pi : G \to U(X)$ of G on X. It is proved in this chapter that $L_{F,\varphi,\psi} : X \to X$ is in S_1 and a formula for the trace of $L_{F,\varphi,\psi} : X \to X$ is given. These results extend, respectively, the corresponding results in Chapter 12 and Chapter 13 from the one-wavelet case to the two-wavelet case. We also give in this chapter the trace class norm inequalities for the localization operator $L_{F,\varphi,\psi} : X \to X$. In order to obtain a lower bound for the norm $\|L_{F,\varphi,\psi}\|_{S_1}$ of $L_{F,\varphi,\psi} : X \to X$, we need the formula (9.1), which is an analogue of the resolution of the identity formula (6.3) for two admissible wavelets for an irreducible and square-integrable representation $\pi : G \to U(X)$ of G on X.

Let φ and ψ be two admissible wavelets for an irreducible and square-integrable representation $\pi : G \to U(X)$ of G on X such that the two-wavelet constant $c_{\varphi,\psi}$ defined by (9.2) is nonzero. Then we have

$$(x,y) = \frac{1}{c_{\varphi,\psi}} \int_G (x, \pi(g)\varphi)(\pi(g)\psi, y)d\mu(g), \quad x, y \in X.$$

We call this formula the resolution of the identity formula for the irreducible and square-integrable representation $\pi : G \to U(X)$ of G on X corresponding to the admissible wavelets φ and ψ.

Let φ and ψ be admissible wavelets for the square-integrable representation $\pi : G \to U(X)$ such that $c_{\varphi,\psi} \neq 0$. Let $F \in L^1(G)$. Then we define the localization operator $L_{F,\varphi,\psi} : X \to X$ by

$$(L_{F,\varphi,\psi}x, y) = \frac{1}{c_{\varphi,\psi}} \int_G F(g)(x, \pi(g)\varphi)(\pi(g)\psi, y)d\mu(g), \quad x, y \in X. \tag{16.1}$$

Theorem 16.1 *The localization operator* $L_{F,\varphi,\psi} : X \to X$ *is in the trace class* S_1 *and*

$$\mathrm{tr}(L_{F,\varphi,\psi}) = \frac{(\psi, \varphi)}{c_{\varphi,\psi}} \int_G F(g)d\mu(g).$$

Proof. By (16.1), Schwarz' inequality, the fact that $\pi : G \to U(X)$ is a unitary representation, and $\|\varphi\| = \|\psi\| = 1$, we get

$$|(L_{F,\varphi,\psi}x, y)|$$

$$\leq \frac{1}{|c_{\varphi,\psi}|}\|x\|\,\|y\| \int_G |F(g)|\,\|\pi(g)\varphi\|\,\|\pi(g)\psi\|d\mu(g) = \frac{1}{|c_{\varphi,\psi}|}\|F\|_{L^1(G)}\|x\|\,\|y\|$$

for all x and y in X. Thus, $L_{F,\varphi,\psi} : X \to X$ is a bounded linear operator. Next, let $\{\varphi_k : k = 1, 2, \ldots\}$ be an orthonormal basis for X. Then, by (16.1), Fubini's theorem, Schwarz' inequality, Parseval's identity, the fact that $\pi : G \to U(X)$ is a unitary representation, and $\|\varphi\| = \|\psi\| = 1$, we get

$$
\begin{aligned}
\sum_{k=1}^{\infty}\|L_{F,\varphi,\psi}\varphi_k\|^2 &= \sum_{k=1}^{\infty}(L_{F,\varphi,\psi}\varphi_k, L_{F,\varphi,\psi}\varphi_k) \\
&= \frac{1}{c_{\varphi,\psi}}\sum_{k=1}^{\infty}\int_G F(g)(\varphi_k, \pi(g)\varphi)(\pi(g)\psi, L_{F,\varphi,\psi}\varphi_k)d\mu(g) \\
&= \frac{1}{c_{\varphi,\psi}}\sum_{k=1}^{\infty}\int_G F(g)(\varphi_k, \pi(g)\varphi)(L_{F,\varphi,\psi}^*\pi(g)\psi, \varphi_k)d\mu(g) \\
&= \frac{1}{c_{\varphi,\psi}}\int_G F(g)(L_{F,\varphi,\psi}^*\pi(g)\psi, \pi(g)\varphi)d\mu(g) \\
&\leq \frac{1}{|c_{\varphi,\psi}|}\int_G |F(g)|\|\pi(g)\varphi\|\,\|L_{F,\varphi,\psi}^*\|_*\|\pi(g)\psi\|d\mu(g) \\
&= \frac{1}{|c_{\varphi,\psi}|}\|L_{F,\varphi,\psi}^*\|_*\|F\|_{L^1(G)} < \infty, \qquad\qquad (16.2)
\end{aligned}
$$

where $L_{F,\varphi,\psi}^* : X \to X$ is the adjoint of $L_{F,\varphi,\psi} : X \to X$. So, by (16.2) and Proposition 2.8, $L_{F,\varphi,\psi} : X \to X$ is in S_2 and hence compact. Now, let $\{\varphi_k : k = 1, 2, \ldots\}$ and $\{\psi_k : k = 1, 2, \ldots\}$ be orthonormal sets in X. Then, by Fubini's theorem, Schwarz' inequality, Bessel's identity, the fact that $\pi : G \to U(X)$ is a unitary representation, and $\|\varphi\| = \|\psi\| = 1$, we get

$$
\begin{aligned}
\sum_{k=1}^{\infty}|(L_{F,\varphi,\psi}\varphi_k, \psi_k)| \\
\leq \frac{1}{|c_{\varphi,\psi}|}\sum_{k=1}^{\infty}\int_G |F(g)|\,|(\varphi_k, \pi(g)\varphi)|\,|(\pi(g)\psi, \psi_k)|d\mu(g) \\
\leq \frac{1}{|c_{\varphi,\psi}|}\int_G |F(g)|\left\{\sum_{k=1}^{\infty}|(\varphi_k, \pi(g)\varphi)|^2\right\}^{\frac{1}{2}}\left\{\sum_{k=1}^{\infty}|(\pi(g)\psi, \psi_k)|^2\right\}^{\frac{1}{2}}d\mu(g) \\
\leq \frac{1}{|c_{\varphi,\psi}|}\int_G |F(g)|\,\|\pi(g)\varphi\|\,\|\pi(g)\psi\|d\mu(g) \\
= \frac{1}{|c_{\varphi,\psi}|}\|F\|_{L^1(G)} < \infty. \qquad\qquad (16.3)
\end{aligned}
$$

Thus, by (16.3) and Proposition 2.5, $L_{F,\varphi,\psi} : X \to X$ is in S_1. Finally, let $\{\varphi_k : k = 1, 2, \ldots\}$ be any orthonormal basis for X. Then, by Fubini's theorem, Parseval's identity, the fact that $\pi : G \to U(X)$ is a unitary representation, and $\|\varphi\| = \|\psi\| = 1$, we get

$$
\begin{aligned}
\operatorname{tr}(L_{F,\varphi,\psi}) &= \sum_{k=1}^{\infty} (L_{F,\varphi,\psi}\varphi_k, \varphi_k) \\
&= \frac{1}{c_{\varphi,\psi}} \sum_{k=1}^{\infty} \int_G F(g)(\varphi_k, \pi(g)\varphi)(\pi(g)\psi, \varphi_k)d\mu(g) \\
&= \frac{1}{c_{\varphi,\psi}} \int_G F(g) \sum_{k=1}^{\infty} (\varphi_k, \pi(g)\varphi)(\pi(g)\psi, \varphi_k)d\mu(g) \\
&= \frac{1}{c_{\varphi,\psi}} \int_G F(g)(\pi(g)\psi, \pi(g)\varphi)d\mu(g) \\
&= \frac{(\psi, \varphi)}{c_{\varphi,\psi}} \int_G F(g)d\mu(g),
\end{aligned}
$$

and the proof is complete. $\qquad\qquad\qquad\qquad\qquad\qquad\qquad\qquad\qquad$ $\square$

Let $F_{\varphi,\psi}$ be the function on G defined by

$$F_{\varphi,\psi}(g) = (L_{F,\varphi,\psi}\pi(g)\varphi, \pi(g)\psi), \quad g \in G. \tag{16.4}$$

The function $F_{\varphi,\psi}$ plays in the two-wavelet theory the role of the function F_φ defined by (14.1). It can be interpreted as the correlation of the filtered signals $L_{F,\varphi,\psi}\pi(g)\varphi$, $g \in G$, generated by the admissible wavelet φ with the other family $\pi(g)\psi$, $g \in G$, of signals generated by the admissible wavelet ψ.

We have the following analogue of Theorem 14.1 on the trace class norm inequalities for two-wavelet localization operators.

Theorem 16.2 *Let $F \in L^1(G)$. Then*

$$\|L_{F,\varphi,\psi}\|_{S_1} \leq \frac{1}{|c_{\varphi,\psi}|}\|F\|_{L^1(G)}.$$

If, in addition, the square-integrable representation $\pi : G \to U(X)$ of G on X is irreducible, then

$$\frac{2}{c_\varphi + c_\psi}\|F_{\varphi,\psi}\|_{L^1(G)} \leq \|L_{F,\varphi,\psi}\|_{S_1}.$$

Remark 16.3 To see that $F_{\varphi,\psi} \in L^1(G)$, we use Fubini's theorem, the left invariance of μ, the definition of one-wavelet constants, and the fact that $\pi : G \to U(X)$

is a unitary representation to get

$$\int_G |F_{\varphi,\psi}(g)|d\mu(g)$$

$$= \int_G \left| \frac{1}{c_{\varphi,\psi}} \int_G F(h)(\pi(g)\varphi, \pi(h)\varphi)(\pi(h)\psi, \pi(g)\psi)d\mu(h) \right| d\mu(g)$$

$$\leq \frac{1}{|c_{\varphi,\psi}|} \int_G \int_G |F(h)| \, |(\pi(g)\varphi, \pi(h)\varphi)| \, |(\pi(h)\psi, \pi(g)\psi)| d\mu(h) d\mu(g)$$

$$\leq \frac{1}{2|c_{\varphi,\psi}|} \int_G |F(h)| \left(\int_G |(\pi(g)\varphi, \pi(h)\varphi)|^2 d\mu(g) \right) d\mu(h)$$

$$+ \frac{1}{2|c_{\varphi,\psi}|} \int_G |F(h)| \left(\int_G |(\pi(h)\psi, \pi(g)\psi)|^2 d\mu(g) \right) d\mu(h)$$

$$\leq \frac{c_\varphi}{2|c_{\varphi,\psi}|} \int_G |F(h)|d\mu(h) + \frac{c_\psi}{2|c_{\varphi,\psi}|} \int_G |F(h)|d\mu(h)$$

$$= \frac{c_\varphi + c_\psi}{2|c_{\varphi,\psi}|} \int_G |F(h)|d\mu(h) < \infty.$$

Remark 16.4 If $\varphi = \psi$ and if F is a real-valued and nonnegative function in $L^1(G)$, then $L_{F,\varphi,\psi} : X \to X$ is a positive operator. So, by Proposition 2.7 and Theorem 13.6,

$$\frac{1}{c_{\varphi,\psi}} \int_G F_{\varphi,\psi}(g)d\mu(g) = \|L_{F,\varphi,\psi}\|_{S_1} = \frac{1}{c_\varphi} \int_G F(g)d\mu(g).$$

Thus, the estimates in Theorem 16.2 are sharp.

Proof of Theorem 16.2. By Theorem 16.1, the localization operator $L_{F,\varphi,\psi} : X \to X$ is in S_1. Using the canonical form for compact operators given in Theorem 2.2, we get

$$L_{F,\varphi,\psi}x = \sum_{k=1}^{\infty} s_k(L_{F,\varphi,\psi})(x, u_k)v_k, \quad x \in X, \tag{16.5}$$

where $s_k(L_{F,\varphi,\psi})$, $k = 1, 2, \ldots$, are the positive singular values of $L_{F,\varphi,\psi} : X \to X$, $\{u_k : k = 1, 2, \ldots\}$ is an orthonormal basis for $N(L_{F,\varphi,\psi})^{\perp}$, $\{v_k : k = 1, 2, \ldots\}$ is an orthonormal set in X and the convergence of the series is understood to be in X. By (16.5),

$$\sum_{j=1}^{\infty}(L_{F,\varphi,\psi}u_j, v_j) = \sum_{j=1}^{\infty} s_j(L_{F,\varphi,\psi}). \tag{16.6}$$

So, by (16.6),

$$\|L_{F,\varphi,\psi}\|_{S_1} = \sum_{j=1}^{\infty}(L_{F,\varphi,\psi}u_j, v_j). \tag{16.7}$$

Thus, by (16.1), (16.7), Fubini's theorem, Parseval's identity, Bessel's inequality, Schwarz' inequality, $\|\varphi\| = \|\psi\| = 1$ and the fact that $\pi : G \to U(X)$ is a unitary representation, we get

$$\|L_{F,\varphi,\psi}\|_{S_1} = \left| \sum_{k=1}^{\infty} (L_{F,\varphi,\psi} u_k, v_k) \right|$$

$$\leq \sum_{k=1}^{\infty} \left| \frac{1}{c_{\varphi,\psi}} \int_G F(g)(u_k, \pi(g)\varphi)(\pi(g)\psi, v_k) d\mu(g) \right|$$

$$\leq \sum_{k=1}^{\infty} \frac{1}{|c_{\varphi,\psi}|} \int_G |F(g)| \, |(u_k, \pi(g)\varphi)| \, |(\pi(g)\psi, v_k)| d\mu(g)$$

$$= \frac{1}{|c_{\varphi,\psi}|} \int_G |F(g)| \sum_{k=1}^{\infty} |(u_k, \pi(g)\varphi)||(\pi(g)\psi, v_k)| d\mu(g)$$

$$\leq \frac{1}{|c_{\varphi,\psi}|} \int_G |F(g)| \left\{ \sum_{k=1}^{\infty} |(u_k, \pi(g)\varphi)|^2 \right\}^{\frac{1}{2}} \left\{ \sum_{k=1}^{\infty} |(\pi(g)\psi, v_k)|^2 \right\}^{\frac{1}{2}} d\mu(g)$$

$$\leq \frac{1}{|c_{\varphi,\psi}|} \int_G |F(g)| \|\pi(g)\varphi\| \|\pi(g)\psi\| d\mu(g)$$

$$= \frac{1}{|c_{\varphi,\psi}|} \int_G |F(g)| d\mu(g) = \frac{1}{|c_{\varphi,\psi}|} \|F\|_{L^1(G)}.$$

Also, by (16.4) and (16.5),

$$|F_{\varphi,\psi}(g)| = |(L_{F,\varphi,\psi} \pi(g)\varphi, \pi(g)\psi)|$$

$$= \left| \sum_{k=1}^{\infty} s_k(L_{F,\varphi,\psi})(\pi(g)\varphi, u_k)(v_k, \pi(g)\psi) \right|$$

$$\leq \frac{1}{2} \sum_{k=1}^{\infty} s_k(L_{F,\varphi,\psi})(|(\pi(g)\varphi, u_k)|^2 + |(v_k, \pi(g)\psi)|^2) \qquad (16.8)$$

for all g in G. So, by the irreducibility of the representation $\pi : G \to U(X)$ and hence the resolution of the identity formula in Theorem 6.1, (16.8), Fubini's theorem and the fact that

$$\|u_k\| = \|v_k\| = 1, \quad k = 1, 2, \ldots,$$

we get

$$\int_G |F_{\varphi,\psi}(g)| d\mu(g)$$

$$\leq \frac{1}{2} \sum_{k=1}^{\infty} s_k(L_{F,\varphi,\psi}) \left\{ \int_G |(\pi(g)\varphi, u_k)|^2 d\mu(g) + \int_G |(v_k, \pi(g)\psi)|^2 d\mu(g) \right\}$$

$$\leq \quad \frac{1}{2}(c_\varphi + c_\psi) \sum_{k=1}^{\infty} s_k(L_{F,\varphi,\psi}). \tag{16.9}$$

Thus, by (16.9),

$$\frac{2}{c_\varphi + c_\psi} \int_G |F_{\varphi,\psi}(g)| d\mu(g) \leq \|L_{F,\varphi,\psi}\|_{S_1},$$

and the proof is complete. $\qquad\qquad\qquad\qquad\qquad\qquad\qquad\qquad\qquad\qquad\qquad\square$

17　The Weyl-Heisenberg Group

We show in this chapter that localization operators on the Weyl-Heisenberg group are the same as the linear operators studied by Daubechies in the paper [12] on signal analysis. We begin with a detailed study of the Weyl-Heisenberg group.

Let $\mathbb{R}^n \times \mathbb{R}^n = \{(q,p) : q, p \in \mathbb{R}^n\}$ and let $\mathbb{Z}$ be the set of all integers. Let $(WH)^n = \mathbb{R}^n \times \mathbb{R}^n \times \mathbb{R}/2\pi\mathbb{Z}$. Then we define the binary operation $\cdot$ on $(WH)^n$ by

$$(q_1, p_1, t_1) \cdot (q_2, p_2, t_2) = (q_1 + q_2, p_1 + p_2, t_1 + t_2 + q_1 \cdot p_2) \tag{17.1}$$

for all points (q_1, p_1, t_1) and (q_2, p_2, t_2) in $(WH)^n$, where $q_1 \cdot p_2$ is the Euclidean inner product of q_1 and p_2 in $\mathbb{R}^n$; t_1, t_2 and $t_1 + t_2 + q_1 \cdot p_2$ are cosets in the quotient group $\mathbb{R}/2\pi\mathbb{Z}$ in which the group law is addition modulo 2π. It is easy to prove the following proposition and we omit the proof.

Proposition 17.1 *With respect to the multiplication $\cdot$ defined by (17.1), $(WH)^n$ is a non-abelian group in which $(0,0,0)$ is the identity element and the inverse element of (q, p, t) is $(-q, -p, -t + q \cdot p)$ for all (q, p, t) in $(WH)^n$.*

Remark 17.2 To simplify the notation a little bit, we identify $\mathbb{R}^n \times \mathbb{R}^n$ with $\mathbb{C}^n$. Thus, $(WH)^n = \mathbb{C}^n \times \mathbb{R}/2\pi\mathbb{Z}$, which can also be identified with $\mathbb{C}^n \times [0, 2\pi] = \mathbb{R}^n \times \mathbb{R}^n \times [0, 2\pi]$.

Proposition 17.3 *The Lebesgue measure $dq\,dp\,dt$ on $\mathbb{R}^n \times \mathbb{R}^n \times [0, 2\pi]$ is the left (and right) Haar measure on $(WH)^n$.*

Proof. To prove left invariance, let f be an integrable function on $(WH)^n$. It is helpful to think of f as a function on $\mathbb{R}^n \times \mathbb{R}^n \times \mathbb{R}$ such that $f(q, p, \cdot)$ is a periodic function with period 2π for fixed but arbitrary q and p in $\mathbb{R}^n$. Then for all (z', t') in $(WH)^n$, we get

$$\int_{(WH)^n} f((z', t') \cdot (z, t)) dz\, dt$$
$$= \int_0^{2\pi} \int_{\mathbb{R}^n} \int_{\mathbb{R}^n} f(q' + q,\, p' + p,\, t' + t + q' \cdot p) dq\, dp\, dt$$
$$= \int_{t' + q' \cdot p}^{2\pi + t' + q' \cdot p} \int_{\mathbb{C}^n} f(z, s) dz\, ds$$
$$= \int_0^{2\pi} \int_{\mathbb{C}^n} f(z, s) dz\, ds$$

$$= \int_{(WH)^n} f(z,t)dz\,dt.$$

The proof for right invariance is similar. $\qquad\qquad\qquad\qquad\qquad\qquad$ □

Remark 17.4 With respect to the multiplication $\cdot$ defined by (17.1), the set $(WH)^n$ becomes a locally compact and Hausdorff group on which the left (and right) Haar measure is the Lebesgue measure on $\mathbb{R}^n \times \mathbb{R}^n \times [0, 2\pi]$. We call $(WH)^n$ the Weyl-Heisenberg group. In light of the existence of a left (and right) Haar measure on $(WH)^n$, $(WH)^n$ is unimodular.

Let $\pi : (WH)^n \to U(L^2(\mathbb{R}^n))$ be the mapping defined by

$$(\pi(q,p,t)f)(x) = e^{i(p\cdot x - q\cdot p + t)}f(x-q), \quad x \in \mathbb{R}^n, \tag{17.2}$$

for all points (q, p, t) in $(WH)^n$ and all functions f in $L^2(\mathbb{R}^n)$.

Proposition 17.5 $\pi : (WH)^n \to U(L^2(\mathbb{R}^n))$ *is a representation of* $(WH)^n$ *on* $L^2(\mathbb{R}^n)$.

Proof. Let (q_1, p_1, t_1) and (q_2, p_2, t_2) be points in $(WH)^n$. Then for all functions f in $L^2(\mathbb{R}^n)$, by (17.2),

$$
\begin{aligned}
&(\pi(q_1,p_1,t_1)\pi(q_2,p_2,t_2)f)(x)\\
=\ & e^{i(p_1\cdot x - q_1\cdot p_1 + t_1)}(\pi(q_2,p_2,t_2)f)(x-q_1)\\
=\ & e^{i(p_1\cdot x - q_1\cdot p_1 + t_1)}e^{i(p_2\cdot(x-q_1) - q_2\cdot p_2 + t_2)}f(x-q_2-q_1)\\
=\ & e^{i((p_1+p_2)\cdot x - (p_1+p_2)\cdot q_1 + t_1 + t_2 - q_2\cdot p_2)}f(x-(q_1+q_2))
\end{aligned}
\tag{17.3}
$$

and

$$
\begin{aligned}
&(\pi((q_1,p_1,t_1)\cdot(q_2,p_2,t_2))f)(x)\\
=\ & (\pi(q_1+q_2, p_1+p_2, t_1+t_2+q_1\cdot p_2)f)(x)\\
=\ & e^{i((p_1+p_2)\cdot x - q_1\cdot p_1 - q_2\cdot p_1 - q_2\cdot p_2 + t_1 + t_2)}f(x-(q_1+q_2))
\end{aligned}
\tag{17.4}
$$

for all x in $\mathbb{R}^n$. Hence, by (17.3) and (17.4),

$$\pi(q_1,p_1,t_1)\pi(q_2,p_2,t_2) = \pi((q_1,p_1,t_1)\cdot(q_2,p_2,t_2))$$

for all points (q_1, p_1, t_1) and (q_2, p_2, t_2) in $(WH)^n$.

It remains to prove that $\pi(q,p,t)f \to f$ in $L^2(\mathbb{R}^n)$ as $(q,p,t) \to (0,0,0)$ for all functions f in $L^2(\mathbb{R}^n)$. But

$$
\begin{aligned}
&\|\pi(q,p,t)f - f\|_{L^2(\mathbb{R}^n)}^2\\
=\ & \int_{\mathbb{R}^n} \left| e^{i(p\cdot x - q\cdot p + t)}f(x-q) - f(x) \right|^2 dx\\
=\ & \int_{\mathbb{R}^n} \left| e^{i(p\cdot x - q\cdot p + t)}\{f(x-q) - f(x)\} + \left\{ e^{i(p\cdot x - q\cdot p + t)} - 1 \right\}f(x) \right|^2 dx\\
\leq\ & 2\int_{\mathbb{R}^n} \{|f(x-q) - f(x)|^2 + |(e^{i(p\cdot x - q\cdot p + t)} - 1)f(x)|^2\}dx
\end{aligned}
\tag{17.5}
$$

for all (q, p, t) in $(WH)^n$ and all functions f in $L^2(\mathbb{R}^n)$. By the L^2-continuity of translations,

$$\int_{\mathbb{R}^n} |f(x - q) - f(x)|^2 dx \to 0 \qquad (17.6)$$

as $q \to 0$. For almost all x in $\mathbb{R}^n$,

$$\left| \left(e^{i(p \cdot x - q \cdot p + t)} - 1 \right) f(x) \right|^2 \to 0 \qquad (17.7)$$

as $(q, p, t) \to (0, 0, 0)$, and

$$\left| \left(e^{i(p \cdot x - q \cdot p + t)} - 1 \right) f(x) \right|^2 \le 4|f(x)|^2. \qquad (17.8)$$

Hence, by (17.7), (17.8) and the Lebesgue dominated convergence theorem,

$$\int_{\mathbb{R}^n} \left| \left(e^{i(p \cdot x - q \cdot p + t)} - 1 \right) f(x) \right|^2 dx \to 0 \qquad (17.9)$$

as $(q, p, t) \to (0, 0, 0)$. Hence, by (17.5), (17.6) and (17.9),

$$\|\pi(q, p, t)f - f\|_{L^2(\mathbb{R}^n)} \to 0$$

as $(q, p, t) \to (0, 0, 0)$, and the proof is complete. □

We call $\pi : (WH)^n \to L^2(\mathbb{R}^n)$ the Schrödinger representation of the Weyl-Heisenberg group $(WH)^n$ on $L^2(\mathbb{R}^n)$ and the following theorem gives us all the information that we want to know about it.

Theorem 17.6 *For all functions f and g in $L^2(\mathbb{R}^n)$, we have*

$$\int_{(WH)^n} |(f, \pi(z, t)g)_{L^2(\mathbb{R}^n)}|^2 dz\, dt = (2\pi)^{n+1} \|f\|^2_{L^2(\mathbb{R}^n)} \|g\|^2_{L^2(\mathbb{R}^n)}. \qquad (17.10)$$

Remark 17.7 The proof of Theorem 17.6 requires some basic knowledge of Fourier analysis, which we assume. Standard references include the books by Goldberg [32], Stein and Weiss [88] and Wong [103].

Proof of Theorem 17.6. We begin with the case when both f and g are in the Schwartz space $\mathcal{S}$. If we denote the left-hand side of (17.10) by $I(f, g)$, then

$$
\begin{aligned}
I(f, g) &= \int_0^{2\pi} \int_{\mathbb{R}^n} \int_{\mathbb{R}^n} \left| \int_{\mathbb{R}^n} f(x) e^{-i(p \cdot x - q \cdot p + t)} \overline{g(x - q)} dx \right|^2 dq\, dp\, dt \\
&= 2\pi \int_{\mathbb{R}^n} \int_{\mathbb{R}^n} \left| \int_{\mathbb{R}^n} e^{-ip \cdot x} f(x) \overline{g(x - q)} dx \right|^2 dq\, dp \\
&= (2\pi)^{n+1} \int_{\mathbb{R}^n} \int_{\mathbb{R}^n} \left| (2\pi)^{-\frac{n}{2}} \int_{\mathbb{R}^n} e^{-ip \cdot x} f(x) (T_{-q}\bar{g})(x) dx \right|^2 dq\, dp \\
&= (2\pi)^{n+1} \int_{\mathbb{R}^n} \int_{\mathbb{R}^n} |(fT_{-q}\bar{g})^\wedge(p)|^2 dq\, dp, \qquad (17.11)
\end{aligned}
$$

where $(T_{-q}\bar{g}) = \bar{g}(x-q)$, x, $q \in \mathbb{R}^n$. So, by (17.11), Plancherel's theorem and Fubini's theorem,

$$
\begin{aligned}
I(f,g) &= (2\pi)^{n+1} \int_{\mathbb{R}^n} \int_{\mathbb{R}^n} |f(x)g(x-q)|^2 dx\, dq \\
&= (2\pi)^{n+1} \int_{\mathbb{R}^n} |f(x)|^2 \left(\int_{\mathbb{R}^n} |g(x-q)|^2 dq \right) dx \\
&= (2\pi)^{n+1} \|f\|^2_{L^2(\mathbb{R}^n)} \|g\|^2_{L^2(\mathbb{R}^n)}.
\end{aligned}
$$

Now, for f, $g \in L^2(\mathbb{R}^n)$, let $\{f_k\}_{k=1}^\infty$ and $\{g_k\}_{k=1}^\infty$ be sequences in $\mathcal{S}$ such that $f_k \to f$ in $L^2(\mathbb{R}^n)$ and $g_k \to g$ in $L^2(\mathbb{R}^n)$ as $k \to \infty$. Then, by what we have just shown,

$$
I(f_k, g_k) \to (2\pi)^{n+1} \|f\|_{L^2(\mathbb{R}^n)} \|g\|_{L^2(\mathbb{R}^n)} \tag{17.12}
$$

as $k \to \infty$. Also,

$$
(f_k, \pi(z,t)g_k)_{L^2(\mathbb{R}^n)} \to (f, \pi(z,t)g)_{L^2(\mathbb{R}^n)} \tag{17.13}
$$

for all (z,t) in $(WH)^n$ as $k \to \infty$. Furthermore, for all j and k,

$$
\begin{aligned}
& |(f_j, \pi(x,t)g_j)_{L^2(\mathbb{R}^n)} - (f_k, \pi(z,t)g_k)_{L^2(\mathbb{R}^n)}|^2 \\
={}& |(f_j - f_k, \pi(z,t)g_j)_{L^2(\mathbb{R}^n)} + (f_k, \pi(z,t)(g_j - g_k))_{L^2(\mathbb{R}^n)}|^2 \\
\leq{}& 2|(f_j - f_k, \pi(z,t)g_j)_{L^2(\mathbb{R}^n)}|^2 + 2|(f_k, \pi(z,t)(g_j - g_k))_{L^2(\mathbb{R}^n)}|^2,
\end{aligned}
$$

and hence, using (17.10) for functions in $\mathcal{S}$, we get a positive constant C such that

$$
\begin{aligned}
& \int_{(WH)^n} |(f_j, \pi(z,t)g_j)_{L^2(\mathbb{R}^n)} - (f_k, \pi(z,t)g_k)_{L^2(\mathbb{R}^n)}|^2 dz\, dt \\
\leq{}& C \left(\|f_j - f_k\|^2_{L^2(\mathbb{R}^n)} + \|g_j - g_k\|^2_{L^2(\mathbb{R}^n)} \right) \to 0
\end{aligned}
$$

as j, $k \to \infty$. So,

$$
(f_k, \pi(\cdot,\cdot)g_k)_{L^2(\mathbb{R}^n)} \to h \tag{17.14}
$$

for some h in $L^2((WH)^n)$ as $k \to \infty$. Therefore there exists a subsequence of $\{(f_k, \pi(\cdot,\cdot)g_k)_{L^2(\mathbb{R}^n)}\}_{k=1}^\infty$, again denoted by $\{(f_k, \pi(\cdot,\cdot)g_k)_{L^2(\mathbb{R}^n)}\}_{k=1}^\infty$, such that

$$
(f_k, \pi(\cdot,\cdot)g_k)_{L^2(\mathbb{R}^n)} \to h \tag{17.15}
$$

a.e. on $(WH)^n$ as $k \to \infty$. Thus, by (17.13)–(17.15),

$$
I(f_k, g_k) \to \int_{(WH)^n} |(f, \pi(z,t)g)_{L^2(\mathbb{R}^n)}|^2 dz\, dt \tag{17.16}
$$

as $k \to \infty$. Hence, by (17.12) and (17.16), the proof is complete. $\qquad\square$

Corollary 17.8 $\pi : (WH)^n \to U(L^2(\mathbb{R}^n))$ *is an irreducible and unitary representation of* $(WH)^n$ *on* $L^2(\mathbb{R}^n)$.

Proof. That $\pi : (WH)^n \to U(L^2(\mathbb{R}^n))$ is a unitary representation of $(WH)^n$ on $L^2(\mathbb{R}^n)$ is an immediate consequence of (17.2). Let M be a nonzero and closed subspace of $L^2(\mathbb{R}^n)$ which is invariant with respect to the representation $\pi : (WH)^n \to U(L^2(\mathbb{R}^n))$. Let g be a nonzero function in M. Then

$$\{\pi(z,t)g : (z,t) \in (WH)^n\} \subseteq M. \tag{17.17}$$

Let $f \in L^2(\mathbb{R}^n)$ be such that f is orthogonal to M. Then, by (17.17),

$$(f, \pi(z,t)g)_{L^2(\mathbb{R}^n)} = 0, \quad (z,t) \in (WH)^n. \tag{17.18}$$

Then, by Theorem 17.6 and (17.18),

$$\|f\|_{L^2(\mathbb{R}^n)}\|g\|_{L^2(\mathbb{R}^n)} = 0$$

and hence $f = 0$. So, M is a dense subspace of $L^2(\mathbb{R}^n)$. Since M is also a closed subspace of $L^2(\mathbb{R}^n)$, it follows that $M = L^2(\mathbb{R}^n)$ and the proof is complete. $\square$

Corollary 17.9 $\pi : (WH)^n \to U(L^2(\mathbb{R}^n))$ *is a square-integrable representation of* $(WH)^n$ *on* $L^2(\mathbb{R}^n)$.

Proof. Let φ be any nonzero function in $L^2(\mathbb{R}^n)$. Then, by Theorem 17.6,

$$\int_{(WH)^n} |(\varphi, \pi(z,t)\varphi)_{L^2(\mathbb{R}^n)}|^2 dz dt = (2\pi)^{n+1}\|\varphi\|_{L^2(\mathbb{R}^n)}^4 < \infty, \tag{17.19}$$

and this completes the proof. $\square$

Corollary 17.10 *Every function* φ *in* $L^2(\mathbb{R}^n)$ *with* $\|\varphi\|_{L^2(\mathbb{R}^n)} = 1$ *is an admissible wavelet for the representation* $\pi : (WH)^n \to U(L^2(\mathbb{R}^n))$ *of* $(WH)^n$ *on* $L^2(\mathbb{R}^n)$ *and*

$$c_\varphi = (2\pi)^{n+1}. \tag{17.20}$$

Corollary 17.10 is an immediate consequence of (17.19).

We can now study localization operators on the Weyl-Heisenberg group $(WH)^n$. To this end, let φ be any function in $L^2(\mathbb{R}^n)$ such that $\|\varphi\|_{L^2(\mathbb{R}^n)} = 1$, and let F be any function in $L^1(\mathbb{R}^n \times \mathbb{R}^n) \cup L^\infty(\mathbb{R}^n \times \mathbb{R}^n)$. Let $F^\sharp$ be the function defined on $(WH)^n$ by

$$F^\sharp(q,p,t) = F(q,p), \quad (q,p,t) \in (WH)^n.$$

Then, by (17.2) and (17.20), the localization operator $L_{F^\sharp,\varphi} : L^2(\mathbb{R}^n) \to L^2(\mathbb{R}^n)$ is given by

$$(L_{F^\sharp,\varphi}f, g)_{L^2(\mathbb{R}^n)}$$

$$= \frac{1}{c_\varphi} \int_0^{2\pi} \int_{\mathbb{R}^n} \int_{\mathbb{R}^n} F(q,p)(f, \pi(q,p,t)\varphi)_{L^2(\mathbb{R}^n)}(\pi(q,p,t)\varphi, g)_{L^2(\mathbb{R}^n)} dq\, dp\, dt$$

$$= (2\pi)^{-n} \int_{\mathbb{R}^n} \int_{\mathbb{R}^n} F(q,p)(f, \varphi_{q,p})_{L^2(\mathbb{R}^n)}(\varphi_{q,p}, g)_{L^2(\mathbb{R}^n)} dq\, dp \qquad (17.21)$$

for all functions f and g in $L^2(\mathbb{R}^n)$, where $\varphi_{q,p}$ is the function on $\mathbb{R}^n$ given by

$$\varphi_{q,p}(x) = e^{ip\cdot x}\varphi(x-q), \quad x \in \mathbb{R}^n, \qquad (17.22)$$

for all q and p in $\mathbb{R}^n$. The localization operator $L_{F^\sharp,\varphi} : L^2(\mathbb{R}^n) \to L^2(\mathbb{R}^n)$ is then exactly the same as the linear operator $D_{F,\varphi} : L^2(\mathbb{R}^n) \to L^2(\mathbb{R}^n)$ given by

$$(D_{F,\varphi}f, g)_{L^2(\mathbb{R}^n)} = (2\pi)^{-n} \int_{\mathbb{R}^n} \int_{\mathbb{R}^n} F(q,p)(f, \varphi_{q,p})_{L^2(\mathbb{R}^n)}(\varphi_{q,p}, g)_{L^2(\mathbb{R}^n)} dq\, dp$$

$$(17.23)$$

for all functions f and g in $L^2(\mathbb{R}^n)$, where $\varphi_{q,p}$ is the function defined by (17.22). The linear operator $D_{F,\varphi} : L^2(\mathbb{R}^n) \to L^2(\mathbb{R}^n)$ is the localization operator first studied in the paper [12] by Daubechies in the context of signal analysis, and hence we call $D_{F,\varphi} : L^2(\mathbb{R}^n) \to L^2(\mathbb{R}^n)$ the Daubechies operator associated to the symbol F and the admissible wavelet φ. See also Section 2.8 of the book [13] by Daubechies in this connection. By (17.20), (17.21), (17.23) and Theorem 14.5, we have the following result.

Theorem 17.11 *Let $F \in L^p(\mathbb{R}^n \times \mathbb{R}^n)$, $1 \le p \le \infty$. Then there exists a unique linear operator $D_{F,\varphi} : L^2(\mathbb{R}^n) \to L^2(\mathbb{R}^n)$ in S_p such that*

$$\|D_{F,\varphi}\|_{S_p} \le (2\pi)^{-\frac{n}{p}} \|F\|_{L^p(\mathbb{R}^n \times \mathbb{R}^n)} \qquad (17.24)$$

and, for all functions f and g in $L^2(\mathbb{R}^n)$, $(D_{F,\varphi}f, g)_{L^2(\mathbb{R}^n)}$ is given by (17.23) for all simple functions F on $\mathbb{R}^n \times \mathbb{R}^n$ such that the Lebesgue measure of the set $\{(q,p) \in \mathbb{R}^n \times \mathbb{R}^n : F(q,p) \ne 0\}$ is finite.

Proof. We only need to check the inequality (17.24). But, by (17.20), (17.21), (17.23) and Theorem 14.5,

$$\|D_{F,\varphi}\|_{S_p} = \|L_{F^\sharp,\varphi}\|_{S_p} \le (2\pi)^{-\frac{n+1}{p}} \|F^\sharp\|_{L^p((WH)^n)}. \qquad (17.25)$$

But, by a simple computation,

$$\|F^\sharp\|_{L^p((WH)^n)} = (2\pi)^{\frac{1}{p}} \|F\|_{L^p(\mathbb{R}^n \times \mathbb{R}^n)}. \qquad (17.26)$$

Thus, by (17.25) and (17.26), (17.24) follows. $\qquad\square$

As a sharp contrast to the Weyl-Heisenberg group $(WH)^n$, we end this chapter by showing that the Heisenberg group H^n introduced in Chapter 8 of the book [102] by Wong is one on which every irreducible and unitary representation of H^n on $L^2(\mathbb{R}^n)$ is not square-integrable, or equivalently, the set $AW(\pi)$ of all admissible wavelets for any irreducible and unitary representation $\pi : H^n \to U(L^2(\mathbb{R}^n))$ of H^n on $L^2(\mathbb{R}^n)$ is empty.

The Heisenberg group H^n is the non-abelian group $\mathbb{C}^n \times \mathbb{R}$ in which the group law $\cdot$ is given by

$$(z, t) \cdot (w, s) = (z + w, t + s + 2\mathrm{Im}(z \cdot \bar{w}))$$

for all (z, t) and (w, s) in $\mathbb{C}^n \times \mathbb{R}$, where

$$z \cdot \bar{w} = \sum_{j=1}^{n} z_j \bar{w}_j.$$

The Heisenberg group H^n is a unimodular group on which the left (and right) Haar measure is the Lebesgue measure $dz\, dt$ on $\mathbb{C}^n \times \mathbb{R}$.

According to the Stone-von Neumann theorem, every irreducible and unitary representation $\pi : H^n \to U(L^2(\mathbb{R}^n))$ of H^n on $L^2(\mathbb{R}^n)$ is, up to unitary equivalence, given by

$$(\pi(z, t)f)(x) = e^{i\lambda(q \cdot x + \frac{1}{2}q \cdot p + \frac{1}{4}t)} f(x + p), \quad x \in \mathbb{R}^n, \tag{17.27}$$

for all functions f in $L^2(\mathbb{R}^n)$, where $\lambda \in \mathbb{R}$ and $(z, t) = (q, p, t)$.

Theorem 17.12 *Every irreducible and unitary representation of H^n on $L^2(\mathbb{R}^n)$ is not square-integrable.*

Proof. Let $\pi : H^n \to U(L^2(\mathbb{R}^n))$ be an irreducible and unitary representation of H^n on $L^2(\mathbb{R}^n)$. Suppose that $\pi : H^n \to U(L^2(\mathbb{R}^n))$ is given by (17.27). Then for all φ in $L^2(\mathbb{R}^n)$,

$$\int_{-\infty}^{\infty} \int_{\mathbb{C}^n} |(\varphi, \pi(z, t)\varphi)_{L^2(\mathbb{R}^n)}|^2 dz\, dt = \int_{-\infty}^{\infty'} \left(\int_{\mathbb{C}^n} |(\varphi, \varphi_{\lambda, q, p})_{L^2(\mathbb{R}^n)}|^2 dq\, dp \right) dt,$$

where

$$\varphi_{\lambda, q, p}(x) = e^{i\lambda q \cdot x} \varphi(x + p), \quad x \in \mathbb{R}^n.$$

Thus,

$$\int_{-\infty}^{\infty} \int_{\mathbb{C}^n} |(\varphi, \pi(z, t)\varphi)_{L^2(\mathbb{R}^n)}|^2 dz\, dt = \infty$$

unless

$$(\varphi, \varphi_{\lambda, q, p})_{L^2(\mathbb{R}^n)} = 0, \quad q, p \in \mathbb{R}^n,$$

or equivalently,

$$\int_{-\infty}^{\infty} \varphi(x)e^{-i\lambda q \cdot x}\bar{\varphi}(x+p)dx = 0, \quad q, p \in \mathbb{R}^n. \tag{17.28}$$

But (17.28) is valid if and only if

$$\varphi(x)\bar{\varphi}(x+p) = 0$$

for almost all x and p in $\mathbb{R}^n$. Thus, $\varphi(x) = 0$ for almost all x in $\mathbb{R}^n$. Indeed, if $\varphi(x) \neq 0$ for all x in a set S with positive measure. Then for all x in S, $\varphi(x+p) = 0$ for almost all p in $\mathbb{R}^n$, and this is a contradiction. Hence the representation $\pi : H^n \to U(L^2(\mathbb{R}^n))$ of H^n on $L^2(\mathbb{R}^n)$ is not square-integrable. $\qquad\square$

18   The Affine Group

We study in this chapter the affine group U, the Hardy space $H_+^2(\mathbb{R})$, and an irreducible and unitary representation $\pi : U \to U(H_+^2(\mathbb{R}))$ of U on $H_+^2(\mathbb{R})$ for which the set $AW(\pi)$ of all admissible wavelets for the representation $\pi : U \to U(H_+^2(\mathbb{R}))$ is a proper subset of the unit sphere with center at the origin in $H_+^2(\mathbb{R})$.

Let U be the upper half plane given by

$$U = \{(b,a) : b \in \mathbb{R}, \ a > 0\}.$$

Then we define the binary operation $\cdot$ on U by

$$(b_1, a_1) \cdot (b_2, a_2) = (b_1 + a_1 b_2, a_1 a_2) \tag{18.1}$$

for all points (b_1, a_1) and (b_2, a_2) in U.

Proposition 18.1 *With respect to the multiplication $\cdot$ defined by (18.1), U is a non-abelian group in which $(0,1)$ is the identity element and the inverse element of (b,a) is $\left(-\frac{b}{a}, \frac{1}{a}\right)$ for all (b,a) in U.*

Proof. Let (b_1, a_1) and (b_2, a_2) be points in U. Then, by (18.1),

$$
\begin{aligned}
((b_1, a_1) \cdot (b_2, a_2)) \cdot (b_3, a_3) &= (b_1 + a_1 b_2, a_1 a_2) \cdot (b_3, a_3) \\
&= (b_1 + a_1 b_2 + a_1 a_2 b_3, a_1 a_2 a_3)
\end{aligned}
$$

and

$$
\begin{aligned}
(b_1, a_1) \cdot ((b_2, a_2) \cdot (b_3, a_3)) &= (b_1, a_1) \cdot (b_2 + a_2 b_3, a_2 a_3) \\
&= (b_1 + a_1 b_2 + a_1 a_2 b_3, a_1 a_2 a_3).
\end{aligned}
$$

Thus, the associative law is valid. For all (b,a) in U, by (18.1),

$$(b,a) \cdot (0,1) = (b,a) \quad \text{and} \quad (0,1) \cdot (b,a) = (b,a).$$

Thus, $(0,1)$ is the identity element. Finally, let $(b,a) \in U$. Then, by (18.1),

$$(b,a) \cdot \left(-\frac{b}{a}, \frac{1}{a}\right) = (0,1) \quad \text{and} \quad \left(-\frac{b}{a}, \frac{1}{a}\right) \cdot (b,a) = (0,1).$$

Hence the inverse element of (b,a) is $\left(-\frac{b}{a}, \frac{1}{a}\right)$. Therefore U is a group with respect to the multiplication $\cdot$ defined by (18.1). That the group U is non-abelian is easy to check and hence omitted. $\qquad\square$

Proposition 18.2 *The left and right Haar measures on U are given by*

$$d\mu = \frac{db\,da}{a^2} \qquad and \qquad dv = \frac{db\,da}{a}$$

respectively.

Proof. To prove left invariance, let f be an integrable function on U with respect to $d\mu$. Then for all (b', a') in U, we get

$$\int_U f((b', a') \cdot (b, a))d\mu = \int_0^\infty \int_{-\infty}^\infty f(b' + a'b, a'a)\frac{db\,da}{a^2}. \qquad (18.2)$$

Let $\beta = b' + a'b$ and $\alpha = a'a$. Then, by (18.2),

$$\int_U f((b', a') \cdot (b, a))d\mu = \int_0^\infty \int_{-\infty}^\infty f(\beta, \alpha)\frac{d\beta\,d\alpha}{\alpha^2} = \int_U f(b, a)d\mu.$$

To prove right invariance, let f be an integrable function on U with respect to dv. Then for all (b', a') in U, we get

$$\int_U f((b, a) \cdot (b', a'))dv = \int_0^\infty \int_{-\infty}^\infty f(b + ab', aa')\frac{db\,da}{a}. \qquad (18.3)$$

Let $\beta = b + ab'$ and $\alpha = aa'$. Then, by (18.3),

$$\int_U f((b, a) \cdot (b', a'))dv = \int_0^\infty \int_{-\infty}^\infty f(\beta, \alpha)\frac{d\beta\,d\alpha}{\alpha} = \int_U f(b, a)dv.$$

$\square$

Remark 18.3 With respect to the multiplication $\cdot$ defined by (18.1), U is a locally compact and Hausdorff group on which the left Haar measure is different from the right Haar measure. Thus, U is a non-unimodular group, which we call the affine group.

Let $H_+^2(\mathbb{R})$ be the subspace of $L^2(\mathbb{R})$ defined by

$$H_+^2(\mathbb{R}) = \{f \in L^2(\mathbb{R}) : \mathrm{supp}(\hat{f}) \subseteq [0, \infty)\},$$

where $\mathrm{supp}(\hat{f})$ is the set of every x in $\mathbb{R}$ for which there is no neighborhood of x on which $\hat{f}$ is equal to zero almost everywhere. Similarly, we define $H_-^2(\mathbb{R})$ to be the subspace of $L^2(\mathbb{R})$ by

$$H_-^2(\mathbb{R}) = \{f \in L^2(\mathbb{R}) : \mathrm{supp}(\hat{f}) \subseteq (-\infty, 0]\}.$$

We call $H_+^2(\mathbb{R})$ and $H_-^2(\mathbb{R})$ the Hardy space and the conjugate Hardy space respectively.

Proposition 18.4 $H_+^2(\mathbb{R})$ and $H_-^2(\mathbb{R})$ are closed subspaces of $L^2(\mathbb{R})$.

Proof. That $H_+^2(\mathbb{R})$ is a subspace of $L^2(\mathbb{R})$ is obvious. Let $\{f_k\}_{k=1}^\infty$ be a sequence in $H_+^2(\mathbb{R})$ such that $f_k \to f$ in $L^2(\mathbb{R})$ as $k \to \infty$. Then, by Plancherel's theorem,

$$\hat{f}_k \to \hat{f}$$

in $L^2(\mathbb{R})$ as $k \to \infty$. Thus, there exists a subsequence of $\{f_k\}_{k=1}^\infty$, again denoted by $\{f_k\}_{k=1}^\infty$, such that

$$\hat{f}_k \to \hat{f} \tag{18.4}$$

a.e. on $\mathbb{R}$ as $k \to \infty$. Using the definition of $H_+^2(\mathbb{R})$ and the definition of $\operatorname{supp}(f_k)$, we get $\hat{f}_k = 0$ a.e. on $(-\infty, 0]$ for $k = 1, 2, \ldots$. Thus, by (18.4), $\hat{f} = 0$ a.e. on $(-\infty, 0]$. Hence $f \in H_+^2(\mathbb{R})$. Therefore $H_+^2(\mathbb{R})$ is a closed subspace of $L^2(\mathbb{R})$. The proof that $H_-^2(\mathbb{R})$ is a closed subspace of $L^2(\mathbb{R})$ is similar. $\qquad\square$

To be specific, only the Hardy space $H_+^2(\mathbb{R})$ is considered. The discussion is equally valid for the conjugate Hardy space $H_-^2(\mathbb{R})$.

Let $\pi : U \to U(H_+^2(\mathbb{R}))$ be the mapping defined by

$$(\pi(b,a)f)(x) = \frac{1}{\sqrt{a}} f\left(\frac{x-b}{a}\right), \quad x \in \mathbb{R}, \tag{18.5}$$

for all points (b, a) in U and all functions f in $H_+^2(\mathbb{R})$.

Proposition 18.5 $\pi : U \to U(H_+^2(\mathbb{R}))$ is a representation of U on $H_+^2(\mathbb{R})$.

To prove Proposition 18.5, we use the subspace W of $H_+^2(\mathbb{R})$ defined by

$$W = \{f \in H_+^2(\mathbb{R}) : \hat{f} \in C_0^\infty(0, \infty)\}.$$

Lemma 18.6 W is a dense subspace of $H_+^2(\mathbb{R})$.

Proof. Let $f \in H_+^2(\mathbb{R})$. Then $\operatorname{supp}(\hat{f}) \subseteq [0, \infty)$. Let $\{\varphi_k\}_{k=1}^\infty$ be a sequence of functions in $C_0^\infty(0, \infty)$ such that

$$\varphi_k \to \hat{f} \tag{18.6}$$

in $L^2(\mathbb{R})$ as $k \to \infty$. For $k = 1, 2, \ldots$, let f_k be the function in $L^2(\mathbb{R})$ such that

$$\hat{f}_k = \varphi_k. \tag{18.7}$$

Then $f_k \in W$, $k = 1, 2, \ldots$, and, by (18.6), (18.7) and Plancherel's theorem, $f_k \to f$ in $L^2(\mathbb{R})$ as $k \to \infty$. Therefore W is a dense subspace of $H_+^2(\mathbb{R})$. $\qquad\square$

Proof of Proposition 18.5. Let (b_1, a_1) and (b_2, a_2) be points in U. Then, by (18.5), we get, for all functions f in $H_+^2(\mathbb{R})$,

$$(\pi(b_1, a_1)\pi(b_2, a_2)f)(x) = \frac{1}{\sqrt{a_1}}(\pi(b_2, a_2)f)\left(\frac{x - b_1}{a_1}\right)$$

$$= \frac{1}{\sqrt{a_1 a_2}}f\left(\frac{x - b_1 - a_1 b_2}{a_1 a_2}\right) \qquad (18.8)$$

and

$$(\pi((b_1, a_1) \cdot (b_2, a_2))f)(x) = (\pi(b_1 + a_1 b_2, a_1 a_2)f)(x)$$

$$= \frac{1}{\sqrt{a_1 a_2}}f\left(\frac{x - b_1 - a_1 b_2}{a_1 a_2}\right) \qquad (18.9)$$

for all x in $\mathbb{R}$. Hence, by (18.8) and (18.9), $\pi : U \to U(H_+^2(\mathbb{R}))$ is a group homomorphism. It remains to prove that $\pi(b, a)f \to f$ in $L^2(\mathbb{R})$ as $(b, a) \to (0, 1)$ for all functions f in $H_+^2(\mathbb{R})$. But, by Plancherel's theorem and the elementary properties of the Fourier transform, we get, for all functions f in W,

$$\|\pi(b, a)f - f\|_{L^2(\mathbb{R})}^2 = \int_{-\infty}^{\infty}\left|\frac{1}{\sqrt{a}}f\left(\frac{x - b}{a}\right) - f(x)\right|^2 dx$$

$$= \int_{-\infty}^{\infty}|\sqrt{a}e^{-ib\xi}\hat{f}(a\xi) - \hat{f}(\xi)|^2 d\xi$$

$$\leq 2\int_{-\infty}^{\infty}|\sqrt{a}e^{-ib\xi}(\hat{f}(a\xi) - \hat{f}(\xi))|^2 d\xi$$

$$+ 2\int_{-\infty}^{\infty}|(\sqrt{a}e^{-ib\xi} - 1)\hat{f}(\xi)|^2 d\xi. \qquad (18.10)$$

For all ξ in $\mathbb{R}$,

$$|(\sqrt{a}e^{-ib\xi} - 1)\hat{f}(\xi)|^2 \to 0 \qquad (18.11)$$

as $(b, a) \to (0, 1)$ and

$$|(\sqrt{a}e^{-ib\xi} - 1)\hat{f}(\xi)|^2 \leq 9|\hat{f}(\xi)|^2 \qquad (18.12)$$

for all b in $\mathbb{R}$ and all a in $(0, 2)$. By (18.11), (18.12) and the Lebesgue dominated convergence theorem,

$$\int_{-\infty}^{\infty}|(\sqrt{a}e^{-ib\xi} - 1)\hat{f}(\xi)|^2 d\xi \to 0 \qquad (18.13)$$

as $(b, a) \to (0, 1)$. For all ξ in $\mathbb{R}$,

$$|\hat{f}(a\xi) - \hat{f}(\xi)|^2 \to 0 \qquad (18.14)$$

as $a \to 1$, and

$$|\hat{f}(a\xi) - \hat{f}(\xi)| \le 2 \sup_{\xi \in \mathbb{R}} |\hat{f}(\xi)| \chi_R(\xi) \qquad (18.15)$$

for all a in $\left(\frac{1}{2}, 2\right)$, where R is a fixed positive number such that

$$\hat{f}(\xi) = 0, \quad \xi > R,$$

and χ_R is the characteristic function on $[0, 2R]$. Thus, by (18.14), (18.15) and the Lebesgue dominated convergence theorem,

$$\int_{-\infty}^{\infty} |\sqrt{a} e^{-ib\xi} (\hat{f}(a\xi) - \hat{f}(\xi))|^2 d\xi \to 0 \qquad (18.16)$$

as $(b, a) \to (0, 1)$. So, by (18.10), (18.13) and (18.16),

$$\pi(b, a)f \to f \qquad (18.17)$$

in $L^2(\mathbb{R})$ as $(b, a) \to (0, 1)$ for all f in W. Let $f \in H_+^2(\mathbb{R})$. Then, by Lemma 18.6, we can find a sequence $\{f_k\}_{k=1}^{\infty}$ of functions in W such that $f_k \to f$ in $L^2(\mathbb{R})$ as $k \to \infty$. Then for any positive number ε, let k_0 be the positive integer such that

$$\|f_{k_0} - f\|_{L^2(\mathbb{R})} < \frac{2\varepsilon}{3}. \qquad (18.18)$$

So, by (18.17), (18.18) and the obvious fact that $\pi(b, a) : H_+^2(\mathbb{R}) \to H_+^2(\mathbb{R})$ is a unitary operator for all (b, a) in U, there exists a positive number δ such that

$$\begin{aligned}
&\|\pi(b, a)f - f\|_{L^2(\mathbb{R})} \\
\le \quad &\|\pi(b, a)(f - f_{k_0})\|_{L^2(\mathbb{R})} + \|\pi(b, a)f_{k_0} - f_{k_0}\|_{L^2(\mathbb{R})} + \|f_{k_0} - f\|_{L^2(\mathbb{R})} < \varepsilon
\end{aligned}$$

whenever (b, a) is within δ-distance of $(0, 1)$. Thus, $\pi(b, a)f \to f$ in $L^2(\mathbb{R})$ for all f in $H_+^2(\mathbb{R})$ as $(b, a) \to (0, 1)$ and the proof is complete. $\qquad \square$

Proposition 18.7 $\pi : U \to U(H_+^2(\mathbb{R}))$ *is an irreducible and unitary representation of U on $H_+^2(\mathbb{R})$.*

Proof. That $\pi(b, a) : H_+^2(\mathbb{R}) \to H_+^2(\mathbb{R})$ is a unitary operator for all (b, a) in U is easy to check and has actually been used in the proof of Proposition 18.5. Let M be a nonzero and closed subspace of $H_+^2(\mathbb{R})$ such that M is invariant with respect to $\pi : U \to U(H_+^2(\mathbb{R}))$. Let g be a nonzero function in M. Then

$$\{\pi(b, a)g : (b, a) \in U\} \subseteq M.$$

Let $f \in H_+^2(\mathbb{R})$ be such that f is orthogonal to M. Then for all points (b, a) in U,

$$\int_{-\infty}^{\infty} f(x) \bar{g} \left(\frac{x - b}{a} \right) dx = 0,$$

and hence, by Plancherel's theorem,

$$\int_{-\infty}^{\infty} e^{ib\xi} \hat{f}(\xi)\overline{\hat{g}(a\xi)}d\xi = 0. \tag{18.19}$$

Thus, by (18.19),

$$\hat{f}(\xi)\overline{\hat{g}(a\xi)} = 0 \tag{18.20}$$

for almost all ξ in $\mathbb{R}$. Suppose that $\hat{f}(\xi) \neq 0$ for all ξ in a set S with positive measure. Then for all ξ in S, by (18.20), we get

$$\hat{g}(a\xi) = 0$$

for all positive numbers a. Thus, $\hat{g} = 0$ and hence $g = 0$. This is a contradiction.

$\square$

To get more information on the irreducible and unitary representation π : $U \to U(H_+^2(\mathbb{R}))$, we need the following subspace A of $H_+^2(\mathbb{R})$ given by

$$A = \left\{ f \in H_+^2(\mathbb{R}) : \int_0^\infty \frac{|\hat{f}(\xi)|^2}{\xi}d\xi < \infty \right\}.$$

Theorem 18.8 *For all f in $H_+^2(\mathbb{R})$ and all g in A,*

$$\int_0^\infty \int_{-\infty}^\infty |(f, \pi(b,a)g)_{L^2(\mathbb{R})}|^2 \frac{db\,da}{a^2} = 2\pi \int_0^\infty |\hat{f}(\xi)|^2 d\xi \int_0^\infty \frac{|\hat{g}(\xi)|^2}{\xi}d\xi.$$

Proof. Let $f \in W$ and $g \in A$. Then, by (18.5), Plancherel's theorem and the elementary properties of the Fourier transform, we get

$$\int_0^\infty \int_{-\infty}^\infty |(f, \pi(b,a)g)_{L^2(\mathbb{R})}|^2 \frac{db\,da}{a^2}$$

$$= \int_0^\infty \int_{-\infty}^\infty \left| \int_{-\infty}^\infty f(x)\bar{g}\left(\frac{x-b}{a}\right)dx \right|^2 \frac{db\,da}{a^3}$$

$$= \int_0^\infty \int_{-\infty}^\infty \left| \int_{-\infty}^\infty e^{ib\xi}\hat{f}(\xi)\overline{\hat{g}(a\xi)}d\xi \right|^2 \frac{db\,da}{a}$$

$$= 2\pi \int_0^\infty \int_{-\infty}^\infty \left| (2\pi)^{-\frac{1}{2}} \int_{-\infty}^\infty e^{ib\xi}\hat{f}(\xi)\overline{\hat{g}(a\xi)}d\xi \right|^2 \frac{db\,da}{a}$$

$$= 2\pi \int_0^\infty \int_{-\infty}^\infty \left| \left(\hat{f}\left(D_a\bar{g}\right)\right)^\vee (b) \right|^2 \frac{db\,da}{a}$$

$$= 2\pi \int_0^\infty \int_{-\infty}^\infty \left| \hat{f}(\xi)\left(D_a\bar{g}\right)(\xi) \right|^2 \frac{d\xi\,da}{a}, \tag{18.21}$$

where

$$\left(D_a\bar{g}\right)(\xi) = \overline{\hat{g}(a\xi)}. \quad \xi \in \mathbb{R}. \tag{18.22}$$

Thus, by (18.21), (18.22) and Fubini's theorem,

$$\int_0^\infty \int_{-\infty}^\infty |(f, \pi(b,a)g)_{L^2(\mathbb{R})}|^2 \frac{db\,da}{a^2}$$

$$= 2\pi \int_0^\infty \left(\int_0^\infty |\hat{f}(\xi)|^2 d\xi \right) |\hat{g}(a\xi)|^2 \frac{da}{a}$$

$$= 2\pi \int_0^\infty |\hat{f}(\xi)|^2 d\xi \int_0^\infty \frac{|\hat{g}(\xi)|^2}{\xi} d\xi \qquad (18.23)$$

for all f in W and all g in A. Now, let $f \in H^2_+(\mathbb{R})$ and $g \in A$. Then, by Lemma 18.6, there exists a sequence $\{f_k\}_{k=1}^\infty$ of functions in W such that

$$f_k \to f \qquad (18.24)$$

in $L^2(\mathbb{R})$ as $k \to \infty$. For $k = 1, 2, \ldots$, we get, by (18.23), (18.24) and Plancherel's theorem,

$$\int_0^\infty \int_{-\infty}^\infty |(f_j, \pi(b,a)g)_{L^2(\mathbb{R})} - (f_k, \pi(b,a)g)_{L^2(\mathbb{R})}|^2 \frac{db\,da}{a^2}$$

$$= 2\pi \int_0^\infty |\hat{f}_j(\xi) - \hat{f}_k(\xi)|^2 d\xi \int_0^\infty \frac{|\hat{g}(\xi)|^2}{\xi} d\xi \to 0$$

as $j, k \to \infty$. So, $\{(f_k, \pi(\cdot, \cdot)g)_{L^2(\mathbb{R})}\}_{k=1}^\infty$ is a Cauchy sequence in $L^2(U)$. Hence there exists a function h in $L^2(U)$ such that

$$(f_k, \pi(\cdot, \cdot)g)_{L^2(\mathbb{R})} \to h \qquad (18.25)$$

in $L^2(U)$ as $k \to \infty$. Therefore there exists a subsequence of $\{(f_k, \pi(\cdot, \cdot)g)_{L^2(\mathbb{R})}\}_{k=1}^\infty$, again denoted by $\{(f_k, \pi(\cdot, \cdot)g)_{L^2(\mathbb{R})}\}_{k=1}^\infty$, such that

$$(f_k, \pi(\cdot, \cdot)g)_{L^2(\mathbb{R})} \to h \qquad (18.26)$$

a.e. on U as $k \to \infty$. But, by (18.24),

$$(f_k, \pi(b,a)g)_{L^2(\mathbb{R})} \to (f, \pi(b,a)g)_{L^2(\mathbb{R})} \qquad (18.27)$$

for all (b, a) in U as $k \to \infty$. So, by (18.25)–(18.27),

$$\int_0^\infty \int_{-\infty}^\infty |(f_k, \pi(b,a)g)_{L^2(\mathbb{R})}|^2 \frac{db\,da}{a^2} \to \int_0^\infty \int_{-\infty}^\infty |(f, \pi(b,a)g)_{L^2(\mathbb{R})}|^2 \frac{db\,da}{a^2}$$

$$(18.28)$$

as $k \to \infty$. But, by (18.23), (18.24) and Plancherel's theorem,

$$\int_0^\infty \int_{-\infty}^\infty |(f_k, \pi(b,a)g)_{L^2(\mathbb{R})}|^2 \frac{db\,da}{a^2} \to 2\pi \int_0^\infty |\hat{f}(\xi)|^2 d\xi \int_0^\infty \frac{|\hat{g}(\xi)|^2}{\xi} d\xi \quad (18.29)$$

as $k \to \infty$. Hence, by (18.28) and (18.29), the proof is complete. $\qquad \square$

Corollary 18.9 $\pi : U \rightarrow U(H^2_+(\mathbb{R}))$ *is a square-integrable representation of U on* $H^2_+(\mathbb{R})$.

Proof. Let $\varphi \in A$. Then, by Theorem 18.8,

$$\int_0^\infty \int_{-\infty}^\infty |(\varphi, \pi(b,a)\varphi)_{L^2(\mathbb{R})}|^2 \frac{db\,da}{a^2}$$
$$= 2\pi \int_0^\infty |\hat{\varphi}(\xi)|^2 d\xi \int_0^\infty \frac{|\hat{\varphi}(\xi)|^2}{\xi} d\xi$$
$$< \infty \tag{18.30}$$

and this completes the proof.   □

Corollary 18.10 *Every function φ in A with $\|\varphi\|_{L^2(\mathbb{R})} = 1$ is an admissible wavelet for the representation $\pi : U \rightarrow U(H^2_+(\mathbb{R}))$ of U on $H^2_+(\mathbb{R})$ and*

$$c_\varphi = 2\pi \int_0^\infty \frac{|\hat{\varphi}(\xi)|^2}{\xi} d\xi.$$

Corollary 18.10 is an immediate consequence of (18.30).

Remark 18.11 Corollary 18.10 tells us that the set $AW(\pi)$ of all admissible wavelets for the representation $\pi : U \rightarrow U(H^2_+(\mathbb{R}))$ of U of $H^2_+(\mathbb{R})$ is nonempty. That $AW(\pi)$ is a proper subset of $\{f \in H^2_+(\mathbb{R}) : \|f\|_{L^2(\mathbb{R})} = 1\}$ is illustrated by the following example.

Example 18.12 Let χ be the characteristic function on $[0,1]$ and let f_0 be the function in $L^2(\mathbb{R})$ such that $\hat{f}_0 = \chi$. Then $f_0 \in H^2_+(\mathbb{R})$. Using the calculations in the derivation of (18.23), we get

$$\int_0^\infty \int_{-\infty}^\infty |(f_0, \pi(b,a)f_0)_{L^2(\mathbb{R})}|^2 \frac{db\,da}{a^2} = \int_0^\infty |\hat{f}_0(\xi)|^2 d\xi \int_0^\infty \frac{|\hat{f}_0(\xi)|^2}{\xi} d\xi. \tag{18.31}$$

But

$$\int_0^\infty \frac{|\hat{f}_0(\xi)|^2}{\xi} d\xi = \int_0^1 \frac{1}{\xi} d\xi = \infty. \tag{18.32}$$

Thus, by (18.31) and (18.32), the function φ on $\mathbb{R}$ defined by

$$\varphi = f_0 / \|f_0\|_{L^2(\mathbb{R})}$$

is in $\{f \in H^2_+(\mathbb{R}) : \|f\|_{L^2(\mathbb{R})} = 1\}$, but not in $AW(\pi)$.

Using Theorem 14.5 and Corollary 18.10, the following result on localization operators on the affine group is immediate.

Theorem 18.13 *Let $\varphi \in A$ and $F \in L^p(U)$, $1 \leq p \leq \infty$. Then the localization operator $L_{F,\varphi} : H_+^2(\mathbb{R}) \to H_+^2(\mathbb{R})$ given by*

$$(L_{F,\varphi}f, g)_{L^2(\mathbb{R})} = \frac{1}{c_\varphi} \int_0^\infty \int_{-\infty}^\infty F(b,a)(f, \pi(b,a)\varphi)_{L^2(\mathbb{R})}(\pi(b,a)\varphi, g)_{L^2(\mathbb{R})} \frac{db\,da}{a^2}$$

for all f and g in $H_+^2(\mathbb{R})$, where

$$c_\varphi = 2\pi \int_0^\infty \frac{|\hat{\varphi}(\xi)|^2}{\xi} d\xi,$$

is in S_p and

$$\|L_{F,\varphi}\|_{S_p} \leq \left(\frac{1}{c_\varphi}\right)^{\frac{1}{p}} \|F\|_{L^p(U)}.$$

19   Wavelet Multipliers

Let $\sigma \in L^{\infty}(\mathbb{R}^n)$. Then we define the linear operator $T_{\sigma} : L^2(\mathbb{R}^n) \to L^2(\mathbb{R}^n)$ by

$$T_{\sigma}u = \mathcal{F}^{-1}\sigma\mathcal{F}u, \quad u \in L^2(\mathbb{R}^n),$$

where $\mathcal{F}$ and $\mathcal{F}^{-1}$ are the Fourier transformation and the inverse Fourier transformation respectively. The Fourier transform $\mathcal{F}u$, sometimes denoted by $\hat{u}$, of a function u in $L^2(\mathbb{R}^n)$ is given by

$$\mathcal{F}u = \lim_{R\to\infty} (\chi_R u)^{\wedge},$$

where χ_R is the characteristic function of the ball with center at the origin and radius R,

$$(\chi_R u)^{\wedge}(\xi) = (2\pi)^{-\frac{n}{2}} \int_{\mathbb{R}^n} e^{-ix\cdot\xi}\chi_R(x)u(x)dx, \quad \xi \in \mathbb{R}^n,$$

and the convergence of $(\chi_R u)^{\wedge}$ to $\mathcal{F}u$ is understood to be in $L^2(\mathbb{R}^n)$. It is a consequence of Plancherel's theorem that $T_{\sigma} : L^2(\mathbb{R}^n) \to L^2(\mathbb{R}^n)$ is a bounded linear operator.

Let φ be any function in $L^2(\mathbb{R}^n) \cap L^4(\mathbb{R}^n) \cap L^{\infty}(\mathbb{R}^n)$ such that $\|\varphi\|_{L^2(\mathbb{R}^n)} = 1$. The aim of this chapter is to make precise the definition of the linear operator $\varphi T_{\sigma}\bar{\varphi} : L^2(\mathbb{R}^n) \to L^2(\mathbb{R}^n)$, where σ is a function in $L^p(\mathbb{R}^n)$, $1 \le p \le \infty$, and to prove that the resulting bounded linear operator is in the Schatten-von Neumann class S_p. To this end, we first prove that if $\sigma \in L^{\infty}(\mathbb{R}^n)$, then the bounded linear operator $\varphi T_{\sigma}\bar{\varphi} : L^2(\mathbb{R}^n) \to L^2(\mathbb{R}^n)$ can be realized as a wavelet multiplier (to be explained) associated to a unitary representation $\pi : \mathbb{R}^n \to U(L^2(\mathbb{R}^n))$ of the additive group $\mathbb{R}^n$ on $L^2(\mathbb{R}^n)$. This connection explains the impetus for the study of the linear operator $\varphi T_{\sigma}\bar{\varphi} : L^2(\mathbb{R}^n) \to L^2(\mathbb{R}^n)$ and also reveals that the techniques developed in Chapters 12–14 can be exploited.

Let $\pi : \mathbb{R}^n \to U(L^2(\mathbb{R}^n))$ be the unitary representation of the additive group $\mathbb{R}^n$ on $L^2(\mathbb{R}^n)$ defined by

$$(\pi(\xi)u)(x) = e^{ix\cdot\xi}u(x), \quad x, \xi \in \mathbb{R}^n, \tag{19.1}$$

for all functions u in $L^2(\mathbb{R}^n)$.

Proposition 19.1 *Let φ be any function in $L^2(\mathbb{R}^n) \cap L^{\infty}(\mathbb{R}^n)$ such that $\|\varphi\|_{L^2(\mathbb{R}^n)} = 1$. Then for all functions u and v in $\mathcal{S}$,*

$$(2\pi)^{-n} \int_{\mathbb{R}^n} (u, \pi(\xi)\varphi)_{L^2(\mathbb{R}^n)}(\pi(\xi)\varphi, v)_{L^2(\mathbb{R}^n)}d\xi = (\varphi u, \varphi v)_{L^2(\mathbb{R}^n)}. \tag{19.2}$$

Proof. Using Plancherel's theorem and the fact that

$$(\pi(\xi)\varphi)^\wedge = T_{-\xi}\hat{\varphi}, \quad \xi \in \mathbb{R}^n,$$

where

$$(T_{-\xi}f)(x) = f(x - \xi), \quad x \in \mathbb{R}^n,$$

for all measurable functions f on $\mathbb{R}^n$, we get

$$(u, \pi(\xi)\varphi)_{L^2(\mathbb{R}^n)} = (\hat{u} * \hat{\psi})(\xi) \tag{19.3}$$

and

$$(\pi(\xi)\varphi, v)_{L^2(\mathbb{R}^n)} = \overline{(\hat{v} * \hat{\psi})(\xi)} \tag{19.4}$$

for all ξ in $\mathbb{R}^n$, where

$$\psi(x) = \overline{\varphi(x)}, \quad x \in \mathbb{R}^n, \tag{19.5}$$

and

$$(\hat{f} * \hat{\psi})(\xi) = \int_{\mathbb{R}^n} \hat{f}(\xi - \eta)\hat{\psi}(\eta)d\eta, \quad \xi \in \mathbb{R}^n,$$

for all functions f in $\mathcal{S}$. Thus, by (19.3)–(19.5), Plancherel's theorem and the fact that

$$(f\psi)^\wedge = (2\pi)^{-\frac{n}{2}}(\hat{f} * \hat{\psi}), \quad f \in \mathcal{S}, \tag{19.6}$$

we get

$$
\begin{aligned}
\int_{\mathbb{R}^n} (u, \pi(\xi)\varphi)_{L^2(\mathbb{R}^n)}(\pi(\xi)\varphi, v)_{L^2(\mathbb{R}^n)}d\xi &= \int_{\mathbb{R}^n} (\hat{u} * \hat{\psi})(\xi)\overline{(\hat{v} * \hat{\psi})(\xi)}d\xi \\
&= (2\pi)^n \int_{\mathbb{R}^n} u(\xi)\psi(\xi)\overline{v(\xi)\psi(\xi)}d\xi \\
&= (2\pi)^n(\varphi u, \varphi v)_{L^2(\mathbb{R}^n)},
\end{aligned}
$$

and the proof is complete. $\qquad\square$

Remark 19.2 Formula (19.2) can be considered as an analogue of the resolution of the identity formula (6.3) for the unitary representation $\pi : \mathbb{R}^n \to U(L^2(\mathbb{R}^n))$ of $\mathbb{R}^n$ on $L^2(\mathbb{R}^n)$.

Proposition 19.3 *Let $\sigma \in L^\infty(\mathbb{R}^n)$ and let φ be any function in $L^2(\mathbb{R}^n) \cap L^\infty(\mathbb{R}^n)$ such that $\|\varphi\|_{L^2(\mathbb{R}^n)} = 1$. If for all functions u in $\mathcal{S}$, we define $P_{\sigma,\varphi}u$ by*

$$(P_{\sigma,\varphi}u, v)_{L^2(\mathbb{R}^n)} = (2\pi)^{-n} \int_{\mathbb{R}^n} \sigma(\xi)(u, \pi(\xi)\varphi)_{L^2(\mathbb{R}^n)}(\pi(\xi)\varphi, v)_{L^2(\mathbb{R}^n)}d\xi \tag{19.7}$$

for all functions v in $\mathcal{S}$, then

$$(P_{\sigma,\varphi}u, v)_{L^2(\mathbb{R}^n)} = ((\varphi T_\sigma \bar{\varphi})u, v)_{L^2(\mathbb{R}^n)}, \quad u, v \in \mathcal{S}.$$

Proof. By (19.3)–(19.5), we get

$$(P_{\sigma,\varphi}u,v)_{L^2(\mathbb{R}^n)} = (2\pi)^{-n}\int_{\mathbb{R}^n}\sigma(\xi)(\hat{u}*\hat{\psi})(\xi)\overline{(\hat{v}*\hat{\psi})(\xi)}d\xi \qquad (19.8)$$

for all functions u and v in $\mathcal{S}$. But, by (19.6) and (19.8),

$$(P_{\sigma,\varphi}u,v)_{L^2(\mathbb{R}^n)} = \int_{\mathbb{R}^n}\sigma(\xi)(u\psi)^{\wedge}(\xi)\overline{(v\psi)^{\wedge}(\xi)}d\xi, \quad u,v\in\mathcal{S}. \qquad (19.9)$$

Thus, by (19.5), (19.9), Plancherel's theorem and the definition of $T_\sigma : L^2(\mathbb{R}^n) \to L^2(\mathbb{R}^n)$, we get

$$
\begin{aligned}
(P_{\sigma,\varphi}u,v)_{L^2(\mathbb{R}^n)} &= (T_\sigma(\psi u),\psi v)_{L^2(\mathbb{R}^n)} \\
&= ((\bar{\psi}T_\sigma\psi)u,v)_{L^2(\mathbb{R}^n)} \\
&= ((\varphi T_\sigma\bar{\varphi})u,v)_{L^2(\mathbb{R}^n)}
\end{aligned}
$$

for all functions u and v in $\mathcal{S}$. $\qquad\square$

Remark 19.4 By Proposition 19.3, the linear operator $\varphi T_\sigma\bar{\varphi} : L^2(\mathbb{R}^n) \to L^2(\mathbb{R}^n)$ associated to σ in $L^\infty(\mathbb{R}^n)$ and φ in $L^2(\mathbb{R}^n)\cap L^\infty(\mathbb{R}^n)$ with the condition that $\|\varphi\|_{L^2(\mathbb{R}^n)} = 1$ is reminiscent of a localization operator studied in Chapter 12. See, in particular, formula (12.1) for the analogy. Had the "admissible wavelet" φ in (19.7) been replaced by the function φ_0 on $\mathbb{R}^n$ given by

$$\varphi_0(x) = 1, \quad x\in\mathbb{R}^n,$$

we would have obtained

$$(P_\sigma u,v)_{L^2(\mathbb{R}^n)} = (T_\sigma u,v)_{L^2(\mathbb{R}^n)}, \quad u,v\in\mathcal{S},$$

i.e., $P_{\sigma,\varphi}$ would have been a "constant coefficient" pseudo-differential operator, or a Fourier multiplier, studied in, say, the book [103] by Wong. In view of the fact that the function φ in the linear operator $\varphi T_\sigma\bar{\varphi} : L^2(\mathbb{R}^n) \to L^2(\mathbb{R}^n)$ plays the role of the admissible wavelet in the linear operator $P_{\sigma,\varphi} : L^2(\mathbb{R}^n) \to L^2(\mathbb{R}^n)$, it is reasonable to call the linear operator $\varphi T_\sigma\bar{\varphi} : L^2(\mathbb{R}^n) \to L^2(\mathbb{R}^n)$ a wavelet multiplier.

In order to define the linear operator $\varphi T_\sigma\bar{\varphi} : L^2(\mathbb{R}^n) \to L^2(\mathbb{R}^n)$, where σ is a function in $L^p(\mathbb{R}^n)$, $1\le p<\infty$, and φ is a function in $L^2(\mathbb{R}^n)\cap L^\infty(\mathbb{R}^n)$ with $\|\varphi\|_{L^2(\mathbb{R}^n)} = 1$, we need some preparation.

Proposition 19.5 *Let $\sigma\in L^1(\mathbb{R}^n)$ and let φ be any function in $L^2(\mathbb{R}^n)\cap L^\infty(\mathbb{R}^n)$ such that $\|\varphi\|_{L^2(\mathbb{R}^n)} = 1$. If for all functions u in $L^2(\mathbb{R}^n)$, we define $P_{\sigma,\varphi}u$ by (19.7) for all functions v in $L^2(\mathbb{R}^n)$, then $P_{\sigma,\varphi} : L^2(\mathbb{R}^n) \to L^2(\mathbb{R}^n)$ is a bounded linear operator and*

$$\|P_{\sigma,\varphi}\|_{B(L^2(\mathbb{R}^n))} \le (2\pi)^{-n}\|\sigma\|_{L^1(\mathbb{R}^n)}, \qquad (19.10)$$

where $\|\ \|_{B(L^2(\mathbb{R}^n))}$ is the norm in the C^*-algebra of all bounded linear operators from $L^2(\mathbb{R}^n)$ into $L^2(\mathbb{R}^n)$.

Proof. By (19.1), (19.7), Schwarz' inequality and the assumption that $\|\varphi\|_{L^2(\mathbb{R}^n)} = 1$,

$$
\begin{aligned}
|(P_{\sigma,\varphi}u, v)_{L^2(\mathbb{R}^n)}| &\leq (2\pi)^{-n} \int_{\mathbb{R}^n} |\sigma(\xi)| |(u, \pi(\xi)\varphi)_{L^2(\mathbb{R}^n)}| |(\pi(\xi)\varphi, v)_{L^2(\mathbb{R}^n)}| d\xi \\
&\leq (2\pi)^{-n} \|\sigma\|_{L^1(\mathbb{R}^n)} \|u\|_{L^2(\mathbb{R}^n)} \|v\|_{L^2(\mathbb{R}^n)}
\end{aligned}
$$

for all functions u and v in $L^2(\mathbb{R}^n)$. $\qquad\square$

Proposition 19.6 *Let $\sigma \in L^p(\mathbb{R}^n)$, $1 < p < \infty$, and let φ be any function in $L^2(\mathbb{R}^n) \cap L^\infty(\mathbb{R}^n)$ such that $\|\varphi\|_{L^2(\mathbb{R}^n)} = 1$. Then there exists a unique bounded linear operator $P_{\sigma,\varphi} : L^2(\mathbb{R}^n) \to L^2(\mathbb{R}^n)$ such that*

$$
\|P_{\sigma,\varphi}\|_{B(L^2(\mathbb{R}^n))} \leq (2\pi)^{-n/p} \|\varphi\|_{L^\infty(\mathbb{R}^n)}^{2/p'} \|\sigma\|_{L^p(\mathbb{R}^n)},
$$

and for all functions u and v in $L^2(\mathbb{R}^n)$, $(P_{\sigma,\varphi}u, v)_{L^2(\mathbb{R}^n)}$ is given by (19.7) for all simple functions σ on $\mathbb{R}^n$ for which the Lebesgue measure of the set $\{\xi \in \mathbb{R}^n : \sigma(\xi) \neq 0\}$ is finite.

Proof. Let $\sigma \in L^\infty(\mathbb{R}^n)$. Then, by (19.1), (19.3)–(19.5), (19.7), Schwarz' inequality and the assumption that $\|\varphi\|_{L^2(\mathbb{R}^n)} = 1$, we get

$$
|(P_\sigma u, v)_{L^2(\mathbb{R}^n)}| \leq (2\pi)^{-n} \|\sigma\|_{L^\infty(\mathbb{R}^n)} \|\hat{u} * \hat{\psi}\|_{L^2(\mathbb{R}^n)} \|\hat{v} * \hat{\psi}\|_{L^2(\mathbb{R}^n)} \qquad (19.11)
$$

for all functions u and v in $L^2(\mathbb{R}^n)$. Using (19.5), (19.6), (19.11) and Plancherel's theorem, we get

$$
\begin{aligned}
|(P_{\sigma,\varphi}u, v)_{L^2(\mathbb{R}^n)}| &\leq \|\sigma\|_{L^\infty(\mathbb{R}^n)} \|u\psi\|_{L^2(\mathbb{R}^n)} \|v\psi\|_{L^2(\mathbb{R}^n)} \\
&\leq \|\sigma\|_{L^\infty(\mathbb{R}^n)} \|\varphi\|_{L^\infty(\mathbb{R}^n)}^2 \|u\|_{L^2(\mathbb{R}^n)} \|v\|_{L^2(\mathbb{R}^n)} \qquad (19.12)
\end{aligned}
$$

for all functions u and v in $L^2(\mathbb{R}^n)$. So, by (19.12),

$$
\|P_{\sigma,\varphi}\|_{B(L^2(\mathbb{R}^n))} \leq \|\varphi\|_{L^\infty(\mathbb{R}^n)}^2 \|\sigma\|_{L^\infty(\mathbb{R}^n)}, \quad \sigma \in L^\infty(\mathbb{R}^n). \qquad (19.13)
$$

Thus, by (19.10), (19.13) and the interpolation argument used in the proof of Proposition 12.3, the proof is complete. $\qquad\square$

Remark 19.7 Propositions 19.5 and 19.6 allow us to define the wavelet multiplier $\varphi T_\sigma \bar{\varphi} : L^2(\mathbb{R}^n) \to L^2(\mathbb{R}^n)$ for all functions σ in $L^p(\mathbb{R}^n)$, $1 \leq p < \infty$, and all functions φ in $L^2(\mathbb{R}^n) \cap L^\infty(\mathbb{R}^n)$ such that $\|\varphi\|_{L^2(\mathbb{R}^n)} = 1$, as the bounded linear operator $P_{\sigma,\varphi} : L^2(\mathbb{R}^n) \to L^2(\mathbb{R}^n)$.

We can now give the Schatten-von Neumann property of the wavelet multiplier $\varphi T_\sigma \bar{\varphi} : L^2(\mathbb{R}^n) \to L^2(\mathbb{R}^n)$, where $\sigma \in L^p(\mathbb{R}^n)$, $1 \leq p \leq \infty$, and φ is a function in $L^2(\mathbb{R}^n) \cap L^\infty(\mathbb{R}^n)$ such that $\|\varphi\|_{L^2(\mathbb{R}^n)} = 1$. The strategy is to look at wavelet multipliers as localization operators associated to symbols σ and admissible wavelets φ. However, not all functions φ in $L^2(\mathbb{R}^n) \cap L^\infty(\mathbb{R}^n)$ with $\|\varphi\|_{L^2(\mathbb{R}^n)} = 1$ can serve as admissible wavelets, which are characterized by the following theorem.

Theorem 19.8 *The set of admissible wavelets for the unitary representation π : $\mathbb{R}^n \to U(L^2(\mathbb{R}^n))$ defined by (19.1) consists of all functions φ in $L^2(\mathbb{R}^n) \cap L^4(\mathbb{R}^n)$ for which $\|\varphi\|_{L^2(\mathbb{R}^n)} = 1$.*

Proof. Let $\varphi \in L^2(\mathbb{R}^n) \cap L^4(\mathbb{R}^n)$ be such that $\|\varphi\|_{L^2(\mathbb{R}^n)} = 1$. Then, by Plancherel's theorem,

$$\int_{\mathbb{R}^n} |(\varphi, \pi(\xi)\varphi)_{L^2(\mathbb{R}^n)}|^2 d\xi = \int_{\mathbb{R}^n} \left| \int_{\mathbb{R}^n} e^{ix \cdot \xi} |\varphi(x)|^2 dx \right|^2 d\xi = (2\pi)^n \|\varphi\|_{L^4(\mathbb{R}^n)}^4.$$

Thus, φ is an admissible wavelet for the unitary representation π of $\mathbb{R}^n$ on $L^2(\mathbb{R}^n)$. The same calculations show that every admissible wavelet for the unitary representation $\pi : \mathbb{R}^n \to U(L^2(\mathbb{R}^n))$ is a function φ in $L^2(\mathbb{R}^n) \cap L^4(\mathbb{R}^n)$ with $\|\varphi\|_{L^2(\mathbb{R}^n)} = 1$. $\square$

Remark 19.9 Let φ be an admissible wavelet for the square-integrable representation $\pi : \mathbb{R}^n \to U(L^2(\mathbb{R}^n))$. Then from the proof of Theorem 19.8, we see that the wavelet constant c_φ is given by

$$c_\varphi = (2\pi)^n \|\varphi\|_{L^4(\mathbb{R}^n)}^4,$$

and hence

$$P_{\sigma,\varphi} = \|\varphi\|_{L^4(\mathbb{R}^n)}^4 L_{\sigma,\varphi},$$

where $L_{\sigma,\varphi} : L^2(\mathbb{R}^n) \to L^2(\mathbb{R}^n)$ is the localization operator on the additive group $\mathbb{R}^n$ associated to the symbol σ and the admissible wavelet φ.

In view of Remark 19.9, we can use Theorem 14.1 to obtain the following theorem.

Theorem 19.10 *Let $\sigma \in L^1(\mathbb{R}^n)$ and let φ be any function in $L^2(\mathbb{R}^n) \cap L^4(\mathbb{R}^n) \cap L^\infty(\mathbb{R}^n)$ such that $\|\varphi\|_{L^2(\mathbb{R}^n)} = 1$. Then the wavelet multiplier $\varphi T_\sigma \bar\varphi : L^2(\mathbb{R}^n) \to L^2(\mathbb{R}^n)$ is in S_1 and*

$$\|\varphi T_\sigma \bar\varphi\|_{S_1} \le (2\pi)^{-n} \|\sigma\|_{L^1(\mathbb{R}^n)}.$$

We are now ready to give the Schatten-von Neumann property of wavelet multipliers.

Theorem 19.11 *Let $\sigma \in L^p(\mathbb{R}^n)$, $1 \le p \le \infty$, and let φ be any function in $L^2(\mathbb{R}^n) \cap L^4(\mathbb{R}^n) \cap L^\infty(\mathbb{R}^n)$ such that $\|\varphi\|_{L^2(\mathbb{R}^n)} = 1$. Then the wavelet multiplier $\varphi T_\sigma \bar\varphi : L^2(\mathbb{R}^n) \to L^2(\mathbb{R}^n)$ is in S_p and*

$$\|\varphi T_\sigma \bar\varphi\|_{S_p} \le \|\varphi\|_{L^\infty(\mathbb{R}^n)}^{\frac{2}{p'}} (2\pi)^{-\frac{n}{p}} \|\sigma\|_{L^p(\mathbb{R}^n)}.$$

Proof. If $p = 1$, then Theorem 19.11 follows from Theorem 19.10. If $p = \infty$, then Theorem 19.11 follows from (19.13). Thus, for $1 < p < \infty$, Theorem 19.11 follows from Theorems 2.10 and 2.11 in the theory of interpolation, and the endpoint cases $p = 1$ and $p = \infty$. $\square$

We end this chapter with a trace formula for wavelet multiplers.

Theorem 19.12 *Let $\sigma \in L^1(\mathbb{R}^n)$ and let φ be any function in $L^2(\mathbb{R}^n) \cap L^4(\mathbb{R}^n) \cap L^\infty(\mathbb{R}^n)$ such that $\|\varphi\|_{L^2(\mathbb{R}^n)} = 1$. Then*

$$\mathrm{tr}(\varphi T_\sigma \bar\varphi) = (2\pi)^{-n} \int_{\mathbb{R}^n} \sigma(\xi) d\xi.$$

Proof. Let $\{\varphi_k : k = 1, 2, \ldots\}$ be an orthonormal basis for $L^2(\mathbb{R}^n)$. Then, using Theorem 19.10, the definition of the trace, Fubini's theorem, Parseval's identity, $\|\varphi\|_{L^2(\mathbb{R}^n)} = 1$ and the fact that $\pi(\xi) : L^2(\mathbb{R}^n) \to L^2(\mathbb{R}^n)$ is a unitary operator for all ξ in $\mathbb{R}^n$, we get

$$
\begin{aligned}
\mathrm{tr}(\varphi T_\sigma \bar\varphi) &= \mathrm{tr}(P_{\sigma,\varphi}) = \sum_{k=1}^{\infty} (P_{\sigma,\varphi}\varphi_k, \varphi_k)_{L^2(\mathbb{R}^n)} \\
&= \sum_{k=1}^{\infty} (2\pi)^{-n} \int_{\mathbb{R}^n} \sigma(\xi) |(\varphi_k, \pi(\xi)\varphi)_{L^2(\mathbb{R}^n)}|^2 d\xi \\
&= (2\pi)^{-n} \int_{\mathbb{R}^n} \sigma(\xi) \sum_{k=1}^{\infty} |(\varphi_k, \pi(\xi)\varphi)_{L^2(\mathbb{R}^n)}|^2 d\xi \\
&= (2\pi)^{-n} \int_{\mathbb{R}^n} \sigma(\xi) \|\pi(\xi)\varphi\|_{L^2(\mathbb{R}^n)}^2 d\xi \\
&= (2\pi)^{-n} \int_{\mathbb{R}^n} \sigma(\xi) d\xi.
\end{aligned}
$$

$\square$

Remark 19.13 It is interesting to note that the trace $\mathrm{tr}(\varphi T_\sigma \bar\varphi)$ of the wavelet multiplier $\varphi T_\sigma \bar\varphi : L^2(\mathbb{R}^n) \to L^2(\mathbb{R}^n)$ is independent of the "admissible wavelet" φ.

20 The Landau-Pollak-Slepian Operator

We show in this chapter that the Landau-Pollak-Slepian operator arising in signal analysis is in fact a wavelet multiplier. We begin with a discussion of the Landau-Pollak-Slepian operator.

Let Ω and T be positive numbers. Then we define the linear operators $P_\Omega :$ $L^2(\mathbb{R}^n) \to L^2(\mathbb{R}^n)$ and $Q_T : L^2(\mathbb{R}^n) \to L^2(\mathbb{R}^n)$ by

$$(P_\Omega f)^\wedge(\xi) = \begin{cases} \hat{f}(\xi), & |\xi| \leq \Omega, \\ 0, & |\xi| > \Omega, \end{cases} \tag{20.1}$$

and

$$(Q_T f)(x) = \begin{cases} f(x), & |x| \leq T, \\ 0, & |x| > T, \end{cases} \tag{20.2}$$

for all functions f in $L^2(\mathbb{R}^n)$.

Proposition 20.1 $P_\Omega : L^2(\mathbb{R}^n) \to L^2(\mathbb{R}^n)$ and $Q_T : L^2(\mathbb{R}^n) \to L^2(\mathbb{R}^n)$ are self-adjoint projections.

Proof. By (20.1) and Plancherel's theorem,

$$
\begin{aligned}
(P_\Omega f, g)_{L^2(\mathbb{R}^n)} &= ((P_\Omega f)^\wedge, \hat{g})_{L^2(\mathbb{R}^n)} = \int_{\mathbb{R}^n} (P_\Omega f)^\wedge(\xi)\overline{\hat{g}(\xi)}d\xi \\
&= \int_{B_\Omega} \hat{f}(\xi)\overline{\hat{g}(\xi)}d\xi = \int_{B_\Omega} \hat{f}(\xi)\overline{(P_\Omega g)^\wedge(\xi)}d\xi \\
&= \int_{\mathbb{R}^n} \hat{f}(\xi)\overline{(P_\Omega g)^\wedge(\xi)}d\xi = (\hat{f}, (P_\Omega g)^\wedge)_{L^2(\mathbb{R}^n)} \\
&= (f, P_\Omega g)_{L^2(\mathbb{R}^n)}
\end{aligned}
$$

for all functions f and g in $L^2(\mathbb{R}^n)$, where B_Ω is the ball in $\mathbb{R}^n$ with center at the origin and radius Ω. Therefore $P_\Omega : L^2(\mathbb{R}^n) \to L^2(\mathbb{R}^n)$ is self-adjoint. By (20.2),

$$
\begin{aligned}
(Q_T f, g)_{L^2(\mathbb{R}^n)} &= \int_{\mathbb{R}^n} (Q_T f)(x)\overline{g(x)}dx = \int_{B_T} f(x)\overline{g(x)}dx \\
&= \int_{\mathbb{R}^n} f(x)\overline{(Q_T g)(x)}dx = (f, Q_T g)_{L^2(\mathbb{R}^n)}
\end{aligned}
$$

for all functions f and g in $L^2(\mathbb{R}^n)$, where B_T is the ball in $\mathbb{R}^n$ with center at the origin and radius T. Therefore $Q_T : L^2(\mathbb{R}^n) \to L^2(\mathbb{R}^n)$ is self-adjoint. By (20.1), the fact that $P_\Omega : L^2(\mathbb{R}^n) \to L^2(\mathbb{R}^n)$ is self-adjoint and Plancherel's theorem,

$$
\begin{aligned}
(P_\Omega^2 f, g)_{L^2(\mathbb{R}^n)} &= (P_\Omega f, P_\Omega g)_{L^2(\mathbb{R}^n)} = ((P_\Omega f)^\wedge, (P_\Omega g)^\wedge)_{L^2(\mathbb{R}^n)} \\
&= \int_{\mathbb{R}^n} (P_\Omega f)^\wedge(\xi)\overline{(P_\Omega g)^\wedge(\xi)}d\xi = \int_{B_\Omega} \hat{f}(\xi)\overline{\hat{g}(\xi)}d\xi \\
&= \int_{\mathbb{R}^n} (P_\Omega f)^\wedge(\xi)\overline{\hat{g}(\xi)}d\xi = ((P_\Omega f)^\wedge, \hat{g})_{L^2(\mathbb{R}^n)} \\
&= (P_\Omega f, g)_{L^2(\mathbb{R}^n)}
\end{aligned}
$$

for all functions f and g in $L^2(\mathbb{R}^n)$. Thus, $P_\Omega^2 = P_\Omega$ and hence $P_\Omega : L^2(\mathbb{R}^n) \to L^2(\mathbb{R}^n)$ is a projection. Finally, by (20.2) and the fact that $P_\Omega : L^2(\mathbb{R}^n) \to L^2(\mathbb{R}^n)$ is self-adjoint,

$$
\begin{aligned}
(Q_T^2 f, g)_{L^2(\mathbb{R}^n)} &= (Q_T f, Q_T g)_{L^2(\mathbb{R}^n)} = \int_{\mathbb{R}^n} (Q_T f)(x)\overline{(Q_T g)(x)}dx \\
&= \int_{B_T} f(x)\overline{g(x)}dx = \int_{\mathbb{R}^n} (Q_T f)(x)\overline{g(x)}dx = (Q_T f, g)_{L^2(\mathbb{R}^n)}
\end{aligned}
$$

for all functions f and g in $L^2(\mathbb{R}^n)$. Thus, $Q_T^2 = Q_T$ and hence $Q_T : L^2(\mathbb{R}^n) \to L^2(\mathbb{R}^n)$ is a projection. $\qquad\square$

In signal analysis, a signal is a function f in $L^2(\mathbb{R}^n)$. Thus, for all functions f in $L^2(\mathbb{R}^n)$, the function $Q_T P_\Omega f$ can be considered to be a time and band-limited signal. Therefore it is of interest to compare the energy $\|Q_T P_\Omega f\|_{L^2(\mathbb{R}^n)}^2$ of the time and band-limited signal $Q_T P_\Omega f$ with the energy $\|f\|_{L^2(\mathbb{R}^n)}^2$ of the original signal f. Using the fact that $P_\Omega : L^2(\mathbb{R}^n) \to L^2(\mathbb{R}^n)$ and $Q_T : L^2(\mathbb{R}^n) \to L^2(\mathbb{R}^n)$ are self-adjoint and the fact that $Q_T : L^2(\mathbb{R}^n) \to L^2(\mathbb{R}^n)$ is a projection, we get

$$
\begin{aligned}
&\sup\left\{\frac{\|Q_T P_\Omega f\|_{L^2(\mathbb{R}^n)}^2}{\|f\|_{L^2(\mathbb{R}^n)}^2} : \quad f \in L^2(\mathbb{R}^n), f \neq 0\right\} \\
&= \sup\left\{\frac{(Q_T P_\Omega f, Q_T P_\Omega f)_{L^2(\mathbb{R}^n)}}{\|f\|_{L^2(\mathbb{R}^n)}^2} : \quad f \in L^2(\mathbb{R}^n), f \neq 0\right\} \\
&= \sup\left\{\frac{(P_\Omega Q_T P_\Omega f, f)_{L^2(\mathbb{R}^n)}}{\|f\|_{L^2(\mathbb{R}^n)}^2} : \quad f \in L^2(\mathbb{R}^n), f \neq 0\right\} \\
&= \sup\left\{(P_\Omega Q_T P_\Omega f, f)_{L^2(\mathbb{R}^n)} : \quad f \in L^2(\mathbb{R}^n), \|f\|_{L^2(\mathbb{R}^n)} = 1\right\}. \quad (20.3)
\end{aligned}
$$

Since $P_\Omega Q_T P_\Omega : L^2(\mathbb{R}^n) \to L^2(\mathbb{R}^n)$ is self-adjoint, it follows from (20.3) that

$$
\sup\left\{\frac{\|Q_T P_\Omega f\|_{L^2(\mathbb{R}^n)}^2}{\|f\|_{L^2(\mathbb{R}^n)}^2} : \quad f \in L^2(\mathbb{R}^n), f \neq 0\right\} = \|P_\Omega Q_T P_\Omega\|_{B(L^2(\mathbb{R}^n))}.
$$

The bounded linear operator $P_\Omega Q_T P_\Omega : L^2(\mathbb{R}^n) \to L^2(\mathbb{R}^n)$ that we have just seen in the context of time and band-limited signals is called the Landau-Pollak-Slepian operator. See the fundamental papers [59, 60] by Landau and Pollak, [83, 84] by Slepian and [85] by Slepian and Pollak for more detailed information.

That the Landau-Pollak-Slepian operator is in fact a wavelet multiplier studied in Chapter 19 is the content of the following theorem.

Theorem 20.2 *Let φ be the function on $\mathbb{R}^n$ defined by*

$$\varphi(x) = \begin{cases} \dfrac{1}{\sqrt{\mu(B_\Omega)}}, & |x| \leq \Omega, \\ 0, & |x| > \Omega, \end{cases} \tag{20.4}$$

where $\mu(B_\Omega)$ is the volume of B_Ω, and let σ be the characteristic function on B_T, i.e.,

$$\sigma(\xi) = \begin{cases} 1, & |\xi| \leq T, \\ 0, & |\xi| > T. \end{cases} \tag{20.5}$$

Then the Landau-Pollak-Slepian operator $P_\Omega Q_T P_\Omega : L^2(\mathbb{R}^n) \to L^2(\mathbb{R}^n)$ is unitarily equivalent to a scalar multiple of the wavelet multiplier $\varphi T_\sigma \varphi : L^2(\mathbb{R}^n) \to L^2(\mathbb{R}^n)$. In fact,

$$P_\Omega Q_T P_\Omega = \mu(B_\Omega) \mathcal{F}^{-1}(\varphi T_\sigma \varphi) \mathcal{F}. \tag{20.6}$$

Proof. By (20.4), φ is a function in $L^2(\mathbb{R}^n) \cap L^\infty(\mathbb{R}^n)$ such that

$$\|\varphi\|_{L^2(\mathbb{R}^n)}^2 = \int_{\mathbb{R}^n} |\varphi(x)|^2 dx = \frac{1}{\mu(B_\Omega)} \int_{B_\Omega} dx = 1.$$

So, by Proposition 19.3,

$$((\varphi T_\sigma \varphi)u, v)_{L^2(\mathbb{R}^n)} = (2\pi)^{-n} \int_{\mathbb{R}^n} \sigma(\xi)(u, \pi(\xi)\varphi)_{L^2(\mathbb{R}^n)} (\pi(\xi)\varphi, v)_{L^2(\mathbb{R}^n)} d\xi \tag{20.7}$$

for all functions u and v in $\mathcal{S}$. By (19.1) and (20.4),

$$\begin{aligned} (u, \pi(\xi)\varphi)_{L^2(\mathbb{R}^n)} &= \int_{\mathbb{R}^n} e^{-ix\cdot\xi} \varphi(x) u(x) dx \\ &= \frac{1}{\sqrt{\mu(B_\Omega)}} \int_{B_\Omega} e^{-ix\cdot\xi} u(x) dx, \quad u \in \mathcal{S}. \end{aligned} \tag{20.8}$$

By (20.1),

$$(P_\Omega \check{u})^\wedge(x) = \begin{cases} u(x), & |x| \leq \Omega, \\ 0, & |x| > \Omega, \end{cases} \tag{20.9}$$

for all functions u in $\mathcal{S}$, where $\check{u}$ is the inverse Fourier transform of u. So, by (20.8), (20.9) and the Fourier inversion formula,

$$
\begin{aligned}
(u, \pi(\xi)\varphi)_{L^2(\mathbb{R}^n)} &= \frac{1}{\sqrt{\mu(B_\Omega)}} \int_{\mathbb{R}^n} e^{-ix\cdot\xi}(P_\Omega \check{u})^\wedge(x)dx \\
&= \frac{(2\pi)^{\frac{n}{2}}}{\sqrt{\mu(B_\Omega)}}(2\pi)^{-\frac{n}{2}} \int_{\mathbb{R}^n} e^{-ix\cdot\xi}(P_\Omega \check{u})^\wedge(x)dx \\
&= \frac{(2\pi)^{\frac{n}{2}}}{\sqrt{\mu(B_\Omega)}}(P_\Omega \check{u})(-\xi), \quad \xi \in \mathbb{R}^n, \quad (20.10)
\end{aligned}
$$

for all functions u in $\mathcal{S}$. Hence, by (20.2), (20.5), (20.7), (20.10), Plancherel's theorem and the fact that $P_\Omega : L^2(\mathbb{R}^n) \to L^2(\mathbb{R}^n)$ is self-adjoint,

$$
\begin{aligned}
((\varphi T_\sigma \varphi)u, v)_{L^2(\mathbb{R}^n)} &= \frac{1}{\mu(B_\Omega)} \int_{\mathbb{R}^n} \sigma(\xi)(P_\Omega \check{u})(\xi)\overline{(P_\Omega \check{v})(\xi)}d\xi \\
&= \frac{1}{\mu(B_\Omega)} \int_{B_T} (P_\Omega \check{u})(\xi)\overline{(P_\Omega \check{v})(\xi)}d\xi \\
&= \frac{1}{\mu(B_\Omega)} \int_{\mathbb{R}^n} (Q_T P_\Omega \check{u})(\xi)\overline{(P_\Omega \check{v})(\xi)}d\xi \\
&= \frac{1}{\mu(B_\Omega)}(Q_T P_\Omega \check{u}, P_\Omega \check{v})_{L^2(\mathbb{R}^n)} \\
&= \frac{1}{\mu(B_\Omega)}(P_\Omega Q_T P_\Omega \check{u}, \check{v})_{L^2(\mathbb{R}^n)} \\
&= \frac{1}{\mu(B_\Omega)}(\mathcal{F} P_\Omega Q_T P_\Omega \mathcal{F}^{-1} u, v)_{L^2(\mathbb{R}^n)}
\end{aligned}
$$

for all functions u and v in $\mathcal{S}$, and hence (20.6) is proved. $\square$

We have the following result on the trace of the Landau-Pollak-Slepian operator $P_\Omega Q_T P_\Omega : L^2(\mathbb{R}^n) \to L^2(\mathbb{R}^n)$.

Theorem 20.3 $\operatorname{tr}(P_\Omega Q_T P_\Omega) = \left\{\frac{n}{2}\Gamma\left(\frac{n}{2}\right)\right\}^{-2}\left(\frac{T\Omega}{2}\right)^n$.

Theorem 20.3 is an immediate consequence of (20.6), Theorem 19.12, Theorem 20.2 and the fact that the volume of the ball in $\mathbb{R}^n$ with radius r is equal to $\frac{\pi^{n/2}r^n}{\frac{n}{2}\Gamma\left(\frac{n}{2}\right)}$.

21   Products of Wavelet Multipliers

The wisdom of the preceding chapter is that a wavelet multiplier can be considered as a filter which time and band-limits a signal. Thus, if we are interested in finding a filter that has the same effect as two wavelet multipliers arranged in series, we are actually seeking a formula for the product of two wavelet multipliers.

We give in this chapter two formulas for the product of two wavelet multipliers $\varphi T_\sigma \bar{\varphi} : L^2(\mathbb{R}^n) \to L^2(\mathbb{R}^n)$ and $\varphi T_\tau \bar{\varphi} : L^2(\mathbb{R}^n) \to L^2(\mathbb{R}^n)$, where σ and τ are functions in $L^2(\mathbb{R}^n)$, and φ is a function in $L^2(\mathbb{R}^n) \cap L^\infty(\mathbb{R}^n)$ such that $\|\varphi\|_{L^2(\mathbb{R}^n)} = 1$. To do this, we need a recall of some basic results without proofs on Weyl transforms from the book [26] by Folland, the books [94, 95] by Thangavelu and the book [102] by Wong.

Let $\sigma \in L^2(\mathbb{R}^n \times \mathbb{R}^n)$. Then the Weyl transform associated to σ is the bounded linear operator $W_\sigma : L^2(\mathbb{R}^n) \to L^2(\mathbb{R}^n)$ given by

$$(W_\sigma f, g)_{L^2(\mathbb{R}^n)} = (2\pi)^{-\frac{n}{2}} \int_{\mathbb{R}^n} \int_{\mathbb{R}^n} \sigma(x, \xi) W(f, g)(x, \xi) dx\, d\xi$$

for all functions f and g in $L^2(\mathbb{R}^n)$, where $W(f, g)$ is the Wigner transform of f and g defined by

$$W(f, g)(x, \xi) = (2\pi)^{-\frac{n}{2}} \int_{\mathbb{R}^n} e^{-i\xi \cdot p} f\left(x + \frac{p}{2}\right) \overline{g\left(x - \frac{p}{2}\right)} dp, \quad x, \xi \in \mathbb{R}^n. \quad (21.1)$$

That the Weyl transform so defined is the same as that defined by (1.1) is well known and can be found in, e.g., Chapter 4 of the book [102] by Wong.

It can be proved that

$$W(f, g)(x, \xi) = V(f, g)^\wedge(x, \xi), \quad x, \xi \in \mathbb{R}^n, \quad (21.2)$$

where the function $V(f, g)$ on $\mathbb{R}^n \times \mathbb{R}^n$ is the Fourier-Wigner transform of f and g defined by

$$V(f, g)(q, p) = (2\pi)^{-\frac{n}{2}} (\rho(q, p)f, g)_{L^2(\mathbb{R}^n)}, \quad q, p \in \mathbb{R}^n, \quad (21.3)$$

and

$$(\rho(q, p)f)(x) = e^{iq \cdot x + \frac{1}{2} iq \cdot p} f(x + p), \quad x \in \mathbb{R}^n. \quad (21.4)$$

For all Schwartz functions f and g on $\mathbb{R}^n$, the functions $V(f, g)$ and $W(f, g)$ are Schwartz functions on $\mathbb{R}^n \times \mathbb{R}^n$.

For all functions f and g in $L^2(\mathbb{R}^n)$, the functions $V(f,g)$ and $W(f,g)$ are in $L^2(\mathbb{R}^n \times \mathbb{R}^n)$. Furthermore, we have

$$\|W(f,g)\|_{L^2(\mathbb{R}^n \times \mathbb{R}^n)} = \|f\|_{L^2(\mathbb{R}^n)}\|g\|_{L^2(\mathbb{R}^n)} \tag{21.5}$$

for all functions f and g in $L^2(\mathbb{R}^n)$. That the same identity is true when W is replaced by V follows from (21.2) and Plancherel's theorem.

Let $h \in L^2(\mathbb{R}^n \times \mathbb{R}^n)$. Then for all functions f in $L^2(\mathbb{R}^n)$, we define the function $K_h f$ on $\mathbb{R}^n$ by

$$(K_h f)(x) = \int_{\mathbb{R}^n} h(x,y)f(y)dy, \quad x \in \mathbb{R}^n.$$

Then $K_h : L^2(\mathbb{R}^n) \to L^2(\mathbb{R}^n)$ is a bounded linear operator and we call it the Hilbert-Schmidt operator corresponding to the kernel h. The following result, obtained by Pool in [69], is the main ingredient in the derivation of the first product formula for two wavelet multipliers.

Proposition 21.1 *Let $h \in L^2(\mathbb{R}^n \times \mathbb{R}^n)$. Then the Hilbert-Schmidt operator corresponding to the kernel h is the same as the Weyl transform $W_\sigma : L^2(\mathbb{R}^n) \to L^2(\mathbb{R}^n)$, and*

$$\sigma = (2\pi)^{\frac{n}{2}} \mathcal{F}_2 Th, \tag{21.6}$$

where $\mathcal{F}_2$ is the Fourier transform on $L^2(\mathbb{R}^n \times \mathbb{R}^n)$ with respect to the second variable and T is the linear operator on $L^2(\mathbb{R}^n \times \mathbb{R}^n)$ defined by

$$(Tf)(x,y) = f\left(x + \frac{y}{2}, x - \frac{y}{2}\right), \quad x, y \in \mathbb{R}^n, \tag{21.7}$$

for all functions f in $L^2(\mathbb{R}^n \times \mathbb{R}^n)$.

We can now give the first formula for the product of two wavelet multipliers.

Theorem 21.2 *Let σ and τ be functions in $L^2(\mathbb{R}^n)$ and let φ be any function in $L^2(\mathbb{R}^n) \cap L^\infty(\mathbb{R}^n)$ such that $\|\varphi\|_{L^2(\mathbb{R}^n)} = 1$. Then the product of the wavelet multipliers $\varphi T_\sigma \bar\varphi : L^2(\mathbb{R}^n) \to L^2(\mathbb{R}^n)$ and $\varphi T_\tau \bar\varphi : L^2(\mathbb{R}^n) \to L^2(\mathbb{R}^n)$ is the same as the linear operator $\varphi W_\lambda \bar\varphi : L^2(\mathbb{R}^n) \to L^2(\mathbb{R}^n)$, where $W_\lambda : L^2(\mathbb{R}^n) \to L^2(\mathbb{R}^n)$ is the Weyl transform associated to λ and*

$$\lambda(x,\xi) = (2\pi)^{-n/2} \int_{\mathbb{R}^n} W(\sigma,\bar\tau)(\xi, y - x)|\varphi(y)|^2 dy \tag{21.8}$$

for all x and ξ in $\mathbb{R}^n$.

The following lemma will be used in the proof of Theorem 21.2.

Lemma 21.3 *Let f and g be functions in $L^2(\mathbb{R}^n)$. Then*

$$W(\check{f}, \check{g})(x,\xi) = W(f,g)(\xi, -x), \quad x, \xi \in \mathbb{R}^n.$$

Proof. Let f and g be functions in $\mathcal{S}$. Then, by (21.3) and (21.4),

$$V(\check{f}, \check{g})(q, p) = (2\pi)^{-\frac{n}{2}} (e^{\frac{1}{2} i q \cdot p} M_q T_p \check{f}, \check{g})_{L^2(\mathbb{R}^n)} \tag{21.9}$$

for all q and p in $\mathbb{R}^n$, where

$$(T_p u)(x) = u(x + p), \quad x \in \mathbb{R}^n,$$

and

$$(M_q u)(x) = e^{i q \cdot x} u(x), \quad x \in \mathbb{R}^n,$$

for all measurable functions u on $\mathbb{R}^n$. So, by (21.3), (21.4), (21.9) and Plancherel's theorem,

$$
\begin{aligned}
V(\check{f}, \check{g})(q, p) &= (2\pi)^{-\frac{n}{2}} (e^{\frac{1}{2} i q \cdot p} (T_{-q} M_p f)^{\vee}, \check{g})_{L^2(\mathbb{R}^n)} \\
&= (2\pi)^{-\frac{n}{2}} (e^{\frac{1}{2} i q \cdot p} T_{-q} M_p f, g)_{L^2(\mathbb{R}^n)} \\
&= (2\pi)^{-\frac{n}{2}} \int_{\mathbb{R}^n} e^{\frac{1}{2} i q \cdot p} e^{i p \cdot (x - q)} f(x - q) \overline{g(x)} dx \\
&= (2\pi)^{-\frac{n}{2}} \int_{\mathbb{R}^n} e^{i p \cdot x - \frac{1}{2} i q \cdot p} f(x - q) \overline{g(x)} dx \\
&= (2\pi)^{-\frac{n}{2}} \int_{\mathbb{R}^n} (\rho(p, -q) f)(x) \overline{g(x)} dx \\
&= V(f, g)(p, -q)
\end{aligned}
\tag{21.10}
$$

for all q and p in $\mathbb{R}^n$. So, by (21.1) and (21.9),

$$
\begin{aligned}
W(\check{f}, \check{g})(x, \xi) &= (2\pi)^{-n} \int_{\mathbb{R}^n} \int_{\mathbb{R}^n} e^{-i q \cdot x - i p \cdot \xi} V(\check{f}, \check{g})(q, p) dq \, dp \\
&= (2\pi)^{-n} \int_{\mathbb{R}^n} \int_{\mathbb{R}^n} e^{-i q \cdot x - i p \cdot \xi} V(f, g)(p, -q) dq \, dp \\
&= (2\pi)^{-n} \int_{\mathbb{R}^n} \int_{\mathbb{R}^n} e^{i q \cdot x - i p \cdot \xi} V(f, g)(p, q) dq \, dp \\
&= W(f, g)(\xi, -x)
\end{aligned}
\tag{21.11}
$$

for all x and ξ in $\mathbb{R}^n$. Thus, by (21.5), (21.10), (21.11), Plancherel's theorem and a limiting argument, the proof is complete. $\square$

Proof of Theorem 21.2. We begin with the observation that for all functions f in $L^1(\mathbb{R}^n) \cap L^2(\mathbb{R}^n)$,

$$(T_\sigma f)(x) = (2\pi)^{-\frac{n}{2}} \int_{\mathbb{R}^n} e^{i x \cdot \xi} \sigma(\xi) \hat{f}(\xi) d\xi = (2\pi)^{-\frac{n}{2}} (\check{\sigma} * f)(x) \tag{21.12}$$

for all x in $\mathbb{R}^n$. Now,

$$(\varphi T_\sigma \bar{\varphi})(\varphi T_\tau \bar{\varphi}) = \varphi T_\sigma \omega T_\tau \bar{\varphi}, \tag{21.13}$$

where

$$\omega = |\varphi|^2, \tag{21.14}$$

and for all functions f in $\mathcal{S}$, we get, by (21.12) and Fubini's theorem,

$$
\begin{aligned}
(T_\sigma \omega T_\tau f)(x) &= (2\pi)^{-\frac{n}{2}} (\check{\sigma} * \omega T_\tau f)(x) \\
&= (2\pi)^{-n} (\check{\sigma} * \omega(\check{\tau} * f))(x) \\
&= (2\pi)^{-n} \int_{\mathbb{R}^n} \check{\sigma}(x-y) \omega(y) (\check{\tau} * f)(y) dy \\
&= (2\pi)^{-n} \int_{\mathbb{R}^n} \check{\sigma}(x-y) \omega(y) \left(\int_{\mathbb{R}^n} \check{\tau}(y-z) f(z) dz \right) dy \\
&= (2\pi)^{-n} \int_{\mathbb{R}^n} \left(\int_{\mathbb{R}^n} \check{\sigma}(x-y) \omega(y) \check{\tau}(y-z) dy \right) f(z) dz \\
&= \int_{\mathbb{R}^n} h(x,z) f(z) dz \qquad (21.15)
\end{aligned}
$$

for all x in $\mathbb{R}^n$, where

$$
h(x,z) = (2\pi)^{-n} \int_{\mathbb{R}^n} \check{\sigma}(x-y) \omega(y) \check{\tau}(y-z) dy, \quad x, z \in \mathbb{R}^n. \qquad (21.16)
$$

By Minkowski's inequality in integral form, Fubini's theorem, Plancherel's theorem and (21.14),

$$
\begin{aligned}
&\left(\int_{\mathbb{R}^n} \int_{\mathbb{R}^n} |h(x,z)|^2 dx\, dz \right)^{\frac{1}{2}} \\
&= (2\pi)^{-n} \left(\int_{\mathbb{R}^n} \int_{\mathbb{R}^n} \left| \int_{\mathbb{R}^n} \check{\sigma}(x-y) \omega(y) \check{\tau}(y-z) dy \right|^2 dx\, dz \right)^{\frac{1}{2}} \\
&\leq (2\pi)^{-n} \int_{\mathbb{R}^n} \left(\int_{\mathbb{R}^n} \int_{\mathbb{R}^n} |\check{\sigma}(x-y) \omega(y) \check{\tau}(y-z)|^2 dx\, dz \right)^{\frac{1}{2}} dy \\
&= (2\pi)^{-n} \int_{\mathbb{R}^n} |\omega(y)| \left(\int_{\mathbb{R}^n} \int_{\mathbb{R}^n} |\check{\sigma}(x-y) \check{\tau}(y-z)|^2 dx\, dz \right)^{\frac{1}{2}} dy \\
&= (2\pi)^{-n} \|\varphi\|_{L^2(\mathbb{R}^n)}^2 \|\sigma\|_{L^2(\mathbb{R}^n)} \|\tau\|_{L^2(\mathbb{R}^n)}. \qquad (21.17)
\end{aligned}
$$

So, by (21.15)–(21.17) and Proposition 21.1, $T_\sigma \omega T_\tau : L^2(\mathbb{R}^n) \to L^2(\mathbb{R}^n)$ is a Weyl transform $W_\lambda : L^2(\mathbb{R}^n) \to L^2(\mathbb{R}^n)$, and by (21.6),

$$
\lambda = (2\pi)^{\frac{n}{2}} \mathcal{F}_2 Th. \qquad (21.18)
$$

By (21.7) and (21.16),

$$
\begin{aligned}
(Th)(x,z) &= (2\pi)^{-n} \int_{\mathbb{R}^n} \check{\sigma} \left(x + \frac{z}{2} - y \right) \omega(y) \check{\tau} \left(y - x + \frac{z}{2} \right) dy \\
&= (2\pi)^{-n} \int_{\mathbb{R}^n} \check{\sigma} \left(x - y + \frac{z}{2} \right) \omega(y) \check{\tau} \left(\frac{z}{2} - (x-y) \right) dy \\
&= (2\pi)^{-n} \int_{\mathbb{R}^n} \check{\sigma} \left(x - y + \frac{z}{2} \right) \omega(y) \overline{\check{\tau} \left(x - y - \frac{z}{2} \right)} dy \qquad (21.19)
\end{aligned}
$$

for all x and z in $\mathbb{R}^n$. For almost all x in $\mathbb{R}^n$, we get, by (21.14), (21.19), Fubini's theorem, Schwarz' inequality and Plancherel's theorem,

$$\int_{\mathbb{R}^n} \left| \int_{\mathbb{R}^n} \breve{\sigma}\left(x - y + \frac{z}{2}\right) \omega(y) \overline{\breve{\tau}\left(x - y - \frac{z}{2}\right)} dy \right| dz$$

$$\leq \int_{\mathbb{R}^n} \left(\int_{\mathbb{R}^n} \left| \breve{\sigma}\left(x - y + \frac{z}{2}\right) \right| |\omega(y)| \left| \breve{\tau}\left(x - y - \frac{z}{2}\right) \right| dy \right) dz$$

$$= \int_{\mathbb{R}^n} |\omega(y)| \left(\int_{\mathbb{R}^n} \left| \breve{\sigma}\left(x - y + \frac{z}{2}\right) \right| \left| \breve{\tau}\left(x - y - \frac{z}{2}\right) \right| dz \right) dy$$

$$\leq \int_{\mathbb{R}^n} |\omega(y)| \left(\int_{\mathbb{R}^n} \left| \breve{\sigma}\left(x - y + \frac{z}{2}\right) \right|^2 dz \right)^{\frac{1}{2}} \left(\int_{\mathbb{R}^n} \left| \breve{\tau}\left(x - y - \frac{z}{2}\right) \right|^2 dz \right)^{\frac{1}{2}} dy$$

$$= 2^n \|\varphi\|_{L^2(\mathbb{R}^n)}^2 \|\sigma\|_{L^2(\mathbb{R}^n)} \|\tau\|_{L^2(\mathbb{R}^n)}.$$

Thus, by (21.1), (21.14), (21.19), Fubini's theorem and Lemma 21.3,

$$(\mathcal{F}_2 Th)(x, \xi)$$

$$= (2\pi)^{-n} \int_{\mathbb{R}^n} \omega(y) \left\{ (2\pi)^{-n/2} \int_{\mathbb{R}^n} e^{-iz\cdot\xi} \breve{\sigma}\left(x - y + \frac{z}{2}\right) \overline{\breve{\tau}\left(x - y - \frac{z}{2}\right)} dz \right\} dy$$

$$= (2\pi)^{-n} \int_{\mathbb{R}^n} \omega(y) W(\breve{\sigma}, \breve{\tau})(x - y, \xi) dy$$

$$= (2\pi)^{-n} \int_{\mathbb{R}^n} W(\sigma, \bar{\tau})(\xi, y - x) \omega(y) dy$$

for all x and ξ in $\mathbb{R}^n$, and hence, by (21.13), (21.14) and (21.18), (21.8) follows. $\square$

In order to give another formula for the product of two wavelet multipliers, we need a recall of a formula, in the paper [35] by Grossmann, Loupias and Stein, for the product of two Weyl transforms associated to functions in $L^2(\mathbb{R}^n \times \mathbb{R}^n)$. To this end, we need the notion of a twisted convolution.

As usual, we identify $\mathbb{R}^n \times \mathbb{R}^n$ with $\mathbb{C}^n$ and any point (q, p) in $\mathbb{R}^n \times \mathbb{R}^n$ with the point $z = q + ip$ in $\mathbb{C}^n$, and we define the symplectic form $[\ ,\ ]$ on $\mathbb{C}^n$ by

$$[z, w] = 2\text{Im}(z \cdot \bar{w}), \quad z, w \in \mathbb{C}^n,$$

where

$$z = (z_1, z_2, \ldots, z_n), \qquad w = (w_1, w_2, \ldots, w_n), \qquad \text{and} \qquad z \cdot \bar{w} = \sum_{j=1}^{n} z_j \bar{w}_j.$$

Now, for any fixed real number λ, we define the twisted convolution $f *_\lambda g$ of two measurable functions f and g on $\mathbb{C}^n$ by

$$(f *_\lambda g)(z) = \int_{\mathbb{C}^n} f(z - w) g(w) e^{i\lambda[z,w]} dw, \quad z \in \mathbb{C}^n, \tag{21.20}$$

where dw is the Lebesgue measure on $\mathbb{C}^n$, provided that the integral exists. The following theorem can be found in the paper [35] by Grossmann, Loupias and Stein.

Theorem 21.4 *Let σ and τ be functions in $L^2(\mathbb{C}^n)$. Then the product of the Weyl transforms $W_\sigma : L^2(\mathbb{R}^n) \to L^2(\mathbb{R}^n)$ and $W_\tau : L^2(\mathbb{R}^n) \to L^2(\mathbb{R}^n)$ is the same as the Weyl transform $W_\omega : L^2(\mathbb{R}^n) \to L^2(\mathbb{R}^n)$, where ω is the function in $L^2(\mathbb{C}^n)$ given by*

$$\hat{\omega} = (2\pi)^{-n}(\hat{\sigma} *_{\frac{1}{4}} \hat{\tau}).$$

A proof of Theorem 21.4 can be found in Chapter 9 of the book [102] by Wong.

Another ingredient in the derivation of another formula for the product of two wavelet multipliers is given in the following theorem.

Theorem 21.5 *Let $\sigma \in L^2(\mathbb{R}^n)$ and let φ be any function in $L^2(\mathbb{R}^n) \cap L^\infty(\mathbb{R}^n)$ such that $\|\varphi\|_{L^2(\mathbb{R}^n)} = 1$. Then the wavelet multiplier $\varphi T_\sigma \bar{\varphi} : L^2(\mathbb{R}^n) \to L^2(\mathbb{R}^n)$ is the same as the Weyl transform $W_{\sigma_\varphi} : L^2(\mathbb{R}^n) \to L^2(\mathbb{R}^n)$, where*

$$\sigma_\varphi(x, \xi) = (2\pi)^{-\frac{n}{2}} \int_{\mathbb{R}^n} W(\varphi, \varphi)(x, \xi - \eta)\sigma(\eta)d\eta, \quad x, \xi \in \mathbb{R}^n. \tag{21.21}$$

Proof. By (21.12), we get, for all functions f in $\mathcal{S}$,

$$
\begin{aligned}
((\varphi T_\sigma \bar{\varphi})f)(x) &= (2\pi)^{-\frac{n}{2}} \varphi(x)(\check{\sigma} * \bar{\varphi}f)(x) \\
&= (2\pi)^{-\frac{n}{2}} \varphi(x) \int_{\mathbb{R}^n} \check{\sigma}(x - y)\bar{\varphi}(y)f(y)dy \\
&= (2\pi)^{-\frac{n}{2}} \int_{\mathbb{R}^n} \varphi(x)\check{\sigma}(x - y)\bar{\varphi}(y)f(y)dy \\
&= \int_{\mathbb{R}^n} h(x, y)f(y)dy
\end{aligned}
\tag{21.22}
$$

for all x in $\mathbb{R}^n$, where

$$h(x, y) = (2\pi)^{-\frac{n}{2}} \varphi(x)\check{\sigma}(x - y)\bar{\varphi}(y), \quad x, y \in \mathbb{R}^n. \tag{21.23}$$

Now, by (21.23), Fubini's theorem and Plancherel's theorem,

$$
\begin{aligned}
&\int_{\mathbb{R}^n} \int_{\mathbb{R}^n} |h(x, y)|^2 dx\, dy \\
&= (2\pi)^{-n} \int_{\mathbb{R}^n} \int_{\mathbb{R}^n} |\varphi(x)|^2 |\check{\sigma}(x - y)|^2 |\varphi(y)|^2 dx\, dy \\
&= (2\pi)^{-n} \int_{\mathbb{R}^n} |\varphi(y)|^2 \left(\int_{\mathbb{R}^n} |\varphi(x)|^2 |\check{\sigma}(x - y)|^2 dx \right) dy \\
&\leq (2\pi)^{-n} \|\varphi\|_{L^\infty(\mathbb{R}^n)}^2 \left(\int_{\mathbb{R}^n} |\varphi(y)|^2 dy \right) \|\check{\sigma}\|_{L^2(\mathbb{R}^n)}^2 \\
&= (2\pi)^{-n} \|\varphi\|_{L^\infty(\mathbb{R}^n)}^2 \|\varphi\|_{L^2(\mathbb{R}^n)}^2 \|\sigma\|_{L^2(\mathbb{R}^n)}^2 < \infty.
\end{aligned}
\tag{21.24}
$$

So, by (21.22)–(21.24), $\varphi T_\sigma \bar{\varphi} : L^2(\mathbb{R}^n) \to L^2(\mathbb{R}^n)$ is a Hilbert-Schmidt operator with kernel h, and hence, by Proposition 21.1, is the same as the Weyl transform $W_{\sigma_\varphi} : L^2(\mathbb{R}^n) \to L^2(\mathbb{R}^n)$, where

$$\sigma_\varphi(x, \xi) = (2\pi)^{\frac{n}{2}} (\mathcal{F}_2 Th)(x, \xi), \quad x, \xi \in \mathbb{R}^n. \tag{21.25}$$

But, by (21.7) and (21.23),

$$\begin{aligned} (Th)(x, y) &= h\left(x + \frac{y}{2}, x - \frac{y}{2}\right) \\ &= (2\pi)^{-\frac{n}{2}} \varphi\left(x + \frac{y}{2}\right) \check{\sigma}(y) \bar{\varphi}\left(x - \frac{y}{2}\right), \quad x, y \in \mathbb{R}^n, \end{aligned}$$

and hence by (19.6),

$$\begin{aligned} (\mathcal{F}_2 Th)(x, \xi) &= (2\pi)^{-n} \int_{\mathbb{R}^n} e^{-i\xi \cdot y} \varphi\left(x + \frac{y}{2}\right) \check{\sigma}(y) \bar{\varphi}\left(x - \frac{y}{2}\right) dy \\ &= (2\pi)^{-n} (W(\varphi, \varphi)(x, \cdot) * \sigma)(\xi) \end{aligned} \tag{21.26}$$

for all x and ξ in $\mathbb{R}^n$. Hence, by (21.25) and (21.26), the proof is complete. $\square$

We can now give another formula for the product of two wavelet multipliers.

Theorem 21.6 *Let σ and τ be functions in $L^2(\mathbb{R}^n)$ and let φ be any function in $L^2(\mathbb{R}^n) \cap L^\infty(\mathbb{R}^n)$ such that $\|\varphi\|_{L^2(\mathbb{R}^n)} = 1$. Then the product of the wavelet multipliers $\varphi T_\sigma \bar{\varphi} : L^2(\mathbb{R}^n) \to L^2(\mathbb{R}^n)$ and $\varphi T_\tau \bar{\varphi} : L^2(\mathbb{R}^n) \to L^2(\mathbb{R}^n)$ is the same as the Weyl transform $W_\lambda : L^2(\mathbb{R}^n) \to L^2(\mathbb{R}^n)$, and λ is the function in $L^2(\mathbb{R}^n \times \mathbb{R}^n)$ given by*

$$\hat{\lambda} = (2\pi)^{-n} (\hat{\sigma}_\varphi *_{\frac{1}{4}} \hat{\tau}_\varphi),$$

where σ_φ and τ_φ are defined by (21.21).

Theorem 21.6 is an immediate consequence of Theorems 21.4 and 21.5.

22 Products of Daubechies Operators

Let $F \in L^2(\mathbb{C}^n)$ and let $\varphi \in L^2(\mathbb{R}^n)$ be such that $\|\varphi\|_{L^2(\mathbb{R}^n)} = 1$. Then the Daubechies operator associated to the symbol F and the admissible wavelet φ is the bounded linear operator $D_{F,\varphi} : L^2(\mathbb{R}^n) \to L^2(\mathbb{R}^n)$ defined by (17.23) for all functions f and g in $L^2(\mathbb{R}^n)$. We give in this chapter a formula for the product of two Daubechies operators when the admissible wavelet φ is chosen to be the function given by

$$\varphi(x) = \pi^{-\frac{n}{4}} e^{-\frac{|x|^2}{2}}, \quad x \in \mathbb{R}^n. \tag{22.1}$$

The starting point is the following theorem.

Theorem 22.1 *Let φ be the function on $\mathbb{R}^n$ given by (22.1) and let Λ be the function on $\mathbb{C}^n$ defined by*

$$\Lambda(z) = \pi^{-n} e^{-|z|^2}, \quad z \in \mathbb{C}^n. \tag{22.2}$$

Then for all functions F in $L^2(\mathbb{C}^n)$, the Daubechies operator $D_{F,\varphi} : L^2(\mathbb{R}^n) \to L^2(\mathbb{R}^n)$ is the Weyl transform $W_{F*\Lambda} : L^2(\mathbb{R}^n) \to L^2(\mathbb{R}^n)$.

Theorem 22.1 is Theorem 17.1 in the book [102] by Wong. It is important to emphasize right at the beginning the fact that the results in this chapter are valid only for the choice of φ defined by (22.1) as the admissible wavelet for the Daubechies operator $D_{F,\varphi} : L^2(\mathbb{R}^n) \to L^2(\mathbb{R}^n)$.

For any fixed real number λ, we define the new twisted convolution or the λ-convolution $f *^\lambda g$ of two measurable functions f and g on $\mathbb{C}^n$ by

$$(f *^\lambda g)(z) = \int_{\mathbb{C}^n} f(z - \omega)g(\omega)e^{\lambda(z \cdot \bar{\omega} - |\omega|^2)} d\omega, \quad z \in \mathbb{C}^n, \tag{22.3}$$

provided that the integral exists. We have the following result.

Theorem 22.2 *Let F and G be in $L^2(\mathbb{C}^n)$. If there exists a function H in $L^2(\mathbb{C}^n)$ such that the Daubechies operator $D_{H,\varphi} : L^2(\mathbb{R}^n) \to L^2(\mathbb{R}^n)$ is the same as the product of the Daubechies operators $D_{F,\varphi} : L^2(\mathbb{R}^n) \to L^2(\mathbb{R}^n)$ and $D_{G,\varphi} : L^2(\mathbb{R}^n) \to L^2(\mathbb{R}^n)$, then*

$$\hat{H} = (2\pi)^{-n}(\hat{F} *^{\frac{1}{2}} \hat{G}). \tag{22.4}$$

Proof. By Theorem 22.1,

$$W_{H*\Lambda} = W_{F*\Lambda} W_{G*\Lambda}. \tag{22.5}$$

It follows from Theorem 21.4 and (22.5) that for all ζ in $\mathbb{C}^n$,

$$(H * \Lambda)^\wedge(\zeta) = (2\pi)^{-n}((F * \Lambda)^\wedge *_{\frac{1}{4}} (G * \Lambda)^\wedge)(\zeta). \qquad (22.6)$$

By (22.2) and an easy computation, we get

$$\hat{\Lambda}(\zeta) = (2\pi)^{-n} e^{-\frac{|\zeta|^2}{4}}, \quad \zeta \in \mathbb{C}^n. \qquad (22.7)$$

Thus, by (22.5)–(22.7) and the definition of the twisted convolution given by (21.20), we get

$$
\begin{aligned}
\hat{H}(\zeta)e^{-\frac{|\zeta|^2}{4}} &= (2\pi)^n \{(\hat{F}\hat{\Lambda}) *_{\frac{1}{4}} (\hat{G}\hat{\Lambda})\}(\zeta) \\
&= (2\pi)^{-n} \int_{\mathbb{C}^n} \hat{F}(\zeta - \omega) e^{-\frac{1}{4}|\zeta - \omega|^2} \hat{G}(\omega) e^{-\frac{1}{4}|\omega|^2} e^{\frac{1}{4}i[\zeta,\omega]} d\omega \\
&= (2\pi)^{-n} \int_{\mathbb{C}^n} \hat{F}(\zeta - \omega) \hat{G}(\omega) e^{\frac{1}{4}\{-|\zeta - \omega|^2 - |\omega|^2 + i[\zeta,\omega]\}} d\omega
\end{aligned}
$$
$$(22.8)$$

for all ζ in $\mathbb{C}^n$. So, by (22.8),

$$\hat{H}(\zeta) = (2\pi)^{-n} \int_{\mathbb{C}^n} \hat{F}(\zeta - \omega)\hat{G}(\omega) e^{\frac{1}{4}\{|\zeta|^2 - |\zeta - \omega|^2 - |\omega|^2 + i[\zeta,\omega]\}} d\omega, \quad \zeta \in \mathbb{C}^n. \quad (22.9)$$

Now, for all ζ and ω in $\mathbb{C}^n$,

$$
\begin{aligned}
&|\zeta|^2 - |\zeta - \omega|^2 - |\omega|^2 + i[\zeta, \omega] \\
&= |\zeta|^2 - |\zeta|^2 + 2\mathrm{Re}(\zeta \cdot \bar{\omega}) - |\omega|^2 - |\omega|^2 + 2i\mathrm{Im}(\zeta \cdot \bar{\omega}) \\
&= 2(\zeta \cdot \bar{\omega}) - 2|\omega|^2.
\end{aligned}
$$
$$(22.10)$$

Therefore, by (22.3), (22.9) and (22.10), we get, for all ζ in $\mathbb{C}^n$,

$$\hat{H}(\zeta) = (2\pi)^{-n} \int_{\mathbb{C}^n} \hat{F}(\zeta - \omega)\hat{G}(\omega) e^{\frac{1}{2}(\zeta \cdot \bar{\omega} - |\omega|^2)} d\omega,$$

and hence (22.4). $\qquad \square$

From the proof of Theorem 22.2, we get the following corollary.

Corollary 22.3 *Let F and G be functions in $L^2(\mathbb{C}^n)$ such that $\hat{F} *_{\frac{1}{2}} \hat{G} \in L^2(\mathbb{C}^n)$. Then there exists a function H in $L^2(\mathbb{C}^n)$ such that $\hat{H} = (2\pi)^{-n} (\hat{F} *_{\frac{1}{2}} \hat{G})$ and the Daubechies operator $D_{H,\varphi} : L^2(\mathbb{R}^n) \to L^2(\mathbb{R}^n)$ is the product of the Daubechies operators $D_{F,\varphi} : L^2(\mathbb{R}^n) \to L^2(\mathbb{R}^n)$ and $D_{G,\varphi} : L^2(\mathbb{R}^n) \to L^2(\mathbb{R}^n)$.*

Remark 22.4 In general, for functions F and G in $L^2(\mathbb{C}^n)$, it is not true that $\hat{F} *_{\frac{1}{2}} \hat{G} \in L^2(\mathbb{C}^n)$. So, the product of two Daubechies operators associated to functions in $L^2(\mathbb{C}^n)$ need not be a Daubechies operator associated to a function in $L^2(\mathbb{C}^n)$. This can best be seen from the following example.

Example 22.5 Let W be the subset of $\mathbb{R} \times \mathbb{R}$ defined by

$$W = \{(q,p) \in \mathbb{R} \times \mathbb{R} : 0 \leq q, p \leq 1\}. \tag{22.11}$$

We identify points ω and ζ in $\mathbb{C}$ with points (q,p) and (x,ξ) in $\mathbb{R} \times \mathbb{R}$ respectively. Let $F \in L^2(\mathbb{C})$ be defined by

$$\hat{F}(q,p) = e^{-\frac{1}{4}|q|}\chi(p), \quad q, p \in \mathbb{R}, \tag{22.12}$$

where χ is the characteristic function on $[-1, 1]$, and let $G \in L^2(\mathbb{C})$ be defined by

$$\hat{G}(\omega) = \begin{cases} e^{\frac{1}{2}|\omega|^2}, & \omega \in W, \\ 0, & \omega \notin W. \end{cases} \tag{22.13}$$

Then, by (22.11)–(22.13),

$$
\begin{aligned}
&(\hat{F} *^{\frac{1}{2}} \hat{G})(\zeta) \\
&= \int_W \hat{F}(\zeta - \omega)\hat{G}(\omega)e^{-\frac{1}{2}|\omega|^2}e^{\frac{1}{2}\zeta\bar{\omega}}\,d\omega \\
&= \int_0^1 \int_0^1 e^{-\frac{1}{4}|x-q|}\chi(\xi - p)e^{\frac{1}{2}(qx+p\xi)}e^{\frac{1}{2}i(q\xi - px)}\,dq\,dp \\
&= \left(\int_0^1 e^{-\frac{1}{4}|x-q|}e^{\frac{1}{2}qx+\frac{1}{2}iq\xi}\,dq \right) \left(\int_0^1 \chi(\xi - p)e^{\frac{1}{2}p\xi - \frac{1}{2}ipx}\,dp \right)
\end{aligned} \tag{22.14}
$$

for all ζ in $\mathbb{C}$. But for $x > 1$ and $0 < \xi < 1$, we get from (22.14)

$$
\begin{aligned}
(\hat{F} *^{\frac{1}{2}} \hat{G})(\zeta) &= \left(\int_0^1 e^{-\frac{1}{4}x}e^{\frac{1}{4}q(1+2\zeta)}\,dq \right) \left(\int_0^1 e^{-\frac{1}{2}ip\zeta}\,dp \right) \\
&= \frac{4e^{-\frac{1}{4}x}}{1+2\zeta} \left(e^{\frac{1}{4}(1+2\zeta)} - 1 \right) \frac{2i}{\zeta} \left(e^{-\frac{1}{2}i\zeta} - 1 \right) \\
&= \frac{4e^{\frac{1}{4}x}}{1+2\zeta} \left(e^{\frac{1}{4}+\frac{1}{2}i\xi} - e^{-\frac{1}{2}x} \right) \frac{2i}{\zeta} \left(e^{\frac{1}{2}\xi}e^{-\frac{1}{2}ix} - 1 \right)
\end{aligned}
$$

and hence $\hat{F} *^{\frac{1}{2}} \hat{G} \notin L^2(\mathbb{C})$.

In view of Remark 22.4 and Example 22.5, it is a natural problem to seek some subspace of $L^2(\mathbb{C}^n)$ such that the product of two Daubechies operators associated to functions in the subspace is indeed a Daubechies operator associated to a function in $L^2(\mathbb{C}^n)$.

For any nonnegative real number c, we denote by $\mathcal{S}_c$ the set of all measurable functions F on $\mathbb{C}^n$ such that

$$|\hat{F}(\zeta)| \leq e^{-c|\zeta|^2}|f(\zeta)|, \quad \zeta \in \mathbb{C}^n,$$

for some function f in $L^2(\mathbb{C}^n)$. It is clear that $\mathcal{S}_c$ is a subspace of $L^2(\mathbb{C}^n)$ for all $c \geq 0$. It is also clear that if $c \leq d$, then $\mathcal{S}_d \subseteq \mathcal{S}_c$.

We can now give a formula for the product of two Daubechies operators associated to functions in $\mathcal{S}_c$, where $c > \frac{1+\sqrt{5}}{8}$.

Theorem 22.6 *Let F and G be functions in $\mathcal{S}_c$, where $c > \frac{1+\sqrt{5}}{8}$. Then the product of the Daubechies operators $D_{F,\varphi} : L^2(\mathbb{R}^n) \to L^2(\mathbb{R}^n)$ and $D_{G,\varphi} : L^2(\mathbb{R}^n) \to L^2(\mathbb{R}^n)$ is the same as the Daubechies operator $D_{H,\varphi} : L^2(\mathbb{R}^n) \to L^2(\mathbb{R}^n)$, where $H \in \bigcap\limits_{0 < d < c'} \mathcal{S}_d$, $c' = c - \frac{1}{4} - \frac{4c^2}{8c+1} > 0$, and*

$$\hat{H} = (2\pi)^{-n}(\hat{F} *^{\frac{1}{2}} \hat{G}).$$

Proof. Let f and g be functions in $L^2(\mathbb{C}^n)$ such that

$$|\hat{F}(\zeta)| \leq e^{-c|\zeta|^2}|f(\zeta)| \tag{22.15}$$

and

$$|\hat{G}(\zeta)| \leq e^{-c|\zeta|^2}|g(\zeta)| \tag{22.16}$$

for all ζ in $\mathbb{C}^n$. Then, by (22.3), (22.15) and (22.16), we get, for all ζ in $\mathbb{C}^n$,

$$
\begin{aligned}
&|(\hat{F} *^{\frac{1}{2}} \hat{G})(\zeta)| \\
&= \left| \int_{\mathbb{C}^n} \hat{F}(\zeta - \omega)\hat{G}(\omega) \, e^{\frac{1}{2}(\zeta \cdot \bar{\omega} - |\omega|^2)} d\omega \right| \\
&\leq \int_{\mathbb{C}^n} |\hat{F}(\zeta - \omega)||\hat{G}(\omega)| \, e^{\frac{1}{2}|\zeta||\omega|} e^{-\frac{1}{2}|\omega|^2} d\omega \\
&\leq \int_{\mathbb{C}^n} e^{-c|\zeta - \omega|^2}|f(\zeta - \omega)| \, e^{-c|\omega|^2}|g(\omega)| \, e^{\frac{1}{4}(|\zeta|^2 + |\omega|^2)} e^{-\frac{1}{2}|\omega|^2} d\omega \\
&\leq e^{-(c - \frac{1}{4})|\zeta|^2} \int_{\mathbb{C}^n} |f(\zeta - \omega)||g(\omega)| \, e^{2c\mathrm{Re}(\zeta \cdot \bar{\omega})} e^{-(2c + \frac{1}{4})|\omega|^2} d\omega. \quad (22.17)
\end{aligned}
$$

But for all positive numbers ε, we have

$$
\begin{aligned}
2c\mathrm{Re}(\zeta \cdot \bar{\omega}) &\leq 2c|\zeta \cdot \bar{\omega}| \leq 2c|\zeta||\omega| \\
&= 2c\sqrt{\varepsilon} |\zeta| \frac{|\omega|}{\sqrt{\varepsilon}} \\
&\leq c\left(\varepsilon|\zeta|^2 + \frac{1}{\varepsilon}|\omega|^2\right) \tag{22.18}
\end{aligned}
$$

for all ζ and ω in $\mathbb{C}^n$. So, by (22.17) and (22.18), we have, for all ζ in $\mathbb{C}^n$,

$$|(\hat{F} *^{\frac{1}{2}} \hat{G})(\zeta)| \leq e^{-(c - \frac{1}{4} - c\varepsilon)|\zeta|^2} \int_{\mathbb{C}^n} |f(\zeta - \omega)||g(\omega)| \, e^{-(2c + \frac{1}{4} - \frac{c}{\varepsilon})|\omega|^2} d\omega. \quad (22.19)$$

Since $c > \frac{1+\sqrt{5}}{8}$, it follows from (22.19) that for any positive number ε such that

$$\frac{c}{2c + \frac{1}{4}} < \varepsilon < 1 - \frac{1}{4c}, \tag{22.20}$$

there exists a positive constant d_ε such that

$$|(\hat{F} *^{\frac{1}{2}} \hat{G})(\zeta)| \le e^{-c_\varepsilon|\zeta|^2} \int_{\mathbb{C}^n} |f(\zeta - \omega)||g(\omega)| e^{-d_\varepsilon|\omega|^2} d\omega, \quad \zeta \in \mathbb{C}^n, \tag{22.21}$$

where

$$c_\varepsilon = c - \frac{1}{4} - c\varepsilon. \tag{22.22}$$

Since, for any ε satisfying (22.20), the function $|g|e^{-d_\varepsilon|\cdot|^2}$ is in $L^1(\mathbb{C}^n)$, it follows from Young's inequality that the function h_ε on $\mathbb{C}^n$ defined by

$$h_\varepsilon(\zeta) = \int_{\mathbb{C}^n} |f(\zeta - \omega)||g(\omega)| e^{-d_\varepsilon|\omega|^2} d\omega, \quad \zeta \in \mathbb{C}^n, \tag{22.23}$$

is in $L^2(\mathbb{C}^n)$. Thus, by (22.21) and (22.23),

$$|(\hat{F} *^{\frac{1}{2}} \hat{G})(\zeta)| \le e^{-c_\varepsilon|\zeta|^2} h_\varepsilon(\zeta), \quad \zeta \in \mathbb{C}^n, \tag{22.24}$$

for any ε satisfying (22.20). Now, by Plancherel's theorem, let $H \in L^2(\mathbb{C}^n)$ be such that

$$\hat{H} = (2\pi)^{-n}(\hat{F} *^{\frac{1}{2}} \hat{G}). \tag{22.25}$$

Then, by (22.24) and (22.25), $H \in S_{c_\varepsilon}$, and hence, by (22.20) and (22.22), $H \in \bigcap_{0<d<c'} S_d$. That the Daubechies operator $D_{H,\varphi} : L^2(\mathbb{R}^n) \to L^2(\mathbb{R}^n)$ is the product of the Daubechies operators $D_{F,\varphi} : L^2(\mathbb{R}^n) \to L^2(\mathbb{R}^n)$ and $D_{G,\varphi} : L^2(\mathbb{R}^n) \to L^2(\mathbb{R}^n)$ is then a consequence of (22.25) and Corollary 22.3. $\qquad\square$

23 Gaussians

As a sequel to the previous chapter, it is of interest to seek another subspace M of $L^2(\mathbb{C}^n)$ such that the product of two Daubechies operators with symbols in M is a Daubechies operator with symbol H in M. We prove in this chapter that M can be taken to be the subspace of $L^2(\mathbb{C}^n)$ spanned by Gaussian functions. Furthermore, an explicit expression for H is given.

Let M be the subspace of $L^2(\mathbb{C}^n)$ spanned by all functions of the form

$$e^{-d|z|^2}, \quad z \in \mathbb{C}^n,$$

where d is a positive number. So, in general, the functions in M are Gaussian functions of the form

$$C_1 e^{-d_1|z|^2} + C_2 e^{-d_2|z|^2} + \cdots + C_m e^{-d_m|z|^2}, \quad z \in \mathbb{C}^n,$$

where C_1, C_2, ..., C_m are complex numbers and d_1, d_2, ..., d_m are positive numbers. Then we have the following theorem.

Theorem 23.1 *Let F and G be functions in M, i.e.,*

$$F(z) = \sum_{k=1}^{m} C_k e^{-d_k|z|^2}, \quad z \in \mathbb{C}^n,$$

and

$$G(z) = \sum_{j=1}^{l} C'_j e^{-d'_j|z|^2}, \quad z \in \mathbb{C}^n,$$

where C_1, C_2, ..., C_m; C'_1, C'_2, ..., C'_l are complex numbers, and d_1, d_2, ..., d_m; d'_1, d'_2, ..., d'_l are positive numbers. Then $D_{F,\varphi} D_{G,\varphi} = D_{H,\varphi}$, where H is also in M and

$$H(z) = \sum_{k=1}^{m} \sum_{j=1}^{l} C_k C'_j e^{-r_{k,j}|z|^2}, \quad z \in \mathbb{C}^n, \tag{23.1}$$

where $r_{k,j} = d_k + d'_j + 2 d_k d'_j$.

To prove Theorem 23.1, we need some preparation. We again identify any points (q, p) and (x, ξ) in $\mathbb{R}^2$ with the points $z = q + ip$ and $\zeta = x + i\xi$ in $\mathbb{C}$ respectively. The following lemma follows from the fact that the Fourier transform of the function ψ given by $\psi(x) = e^{-\frac{x^2}{2}}$, $x \in \mathbb{R}$, is equal to ψ.

Lemma 23.2 $\int_0^\infty e^{-x^2} \cos(2hx)dx = \frac{\sqrt{\pi}}{2} e^{-h^2}$, *where h is any real number.*

Theorem 23.3 *Let*

$$F(z) = e^{-c|z|^2}, \quad z \in \mathbb{C}^n,$$

and

$$G(z) = e^{-d|z|^2}, \quad z \in \mathbb{C}^n,$$

where c and d are positive numbers. Then

$$(F *^{\frac{1}{2}} G)(z) = \left(\frac{\pi}{r}\right)^n e^{-\frac{cd}{r}|z|^2}, \quad z \in \mathbb{C}^n,$$

where

$$r = c + d + \frac{1}{2}. \tag{23.2}$$

Proof. First, we consider the case when $n = 1$. By (22.3) and (23.2), for all z in $\mathbb{C}$,

$$
\begin{aligned}
&(F *^{\frac{1}{2}} G)(z) \\
&= \int_{\mathbb{C}} e^{-c|z-\varsigma|^2} e^{-d|\varsigma|^2} e^{\frac{1}{2}(z\bar\varsigma - |\varsigma|^2)} d\varsigma \\
&= \int_{\mathbb{C}} e^{-c|z|^2 + 2c\mathrm{Re}(z\bar\varsigma) - c|\varsigma|^2} e^{-d|\varsigma|^2} e^{\frac{1}{2}z\bar\varsigma - \frac{1}{2}|\varsigma|^2} d\varsigma \\
&= \int_{\mathbb{C}} e^{-c|z|^2} e^{-(c+d+\frac{1}{2})|\varsigma|^2} e^{2c\mathrm{Re}(z\bar\varsigma)} e^{\frac{1}{2}z\bar\varsigma} d\varsigma \\
&= \int_{\mathbb{R}^2} e^{-c(q^2+p^2)} e^{-(c+d+\frac{1}{2})(x^2+\xi^2)} e^{2c(qx+p\xi)} e^{\frac{1}{2}[qx+p\xi+i(px-q\xi)]} dx\, d\xi \\
&= \int_{\mathbb{R}^2} e^{-c(q^2+p^2)} e^{-r(x^2+\xi^2)} e^{(2c+\frac{1}{2})(qx+p\xi)} e^{i\frac{1}{2}(px-q\xi)} dx\, d\xi \\
&= e^{\left[\frac{(4c+1)^2}{16r} - c\right](q^2+p^2)} \int_{\mathbb{R}^2} e^{-\left(\frac{4c+1}{4\sqrt{r}}q - \sqrt{r}x\right)^2} e^{-\left(\frac{4c+1}{4\sqrt{r}}p - \sqrt{r}\xi\right)^2} e^{i\frac{1}{2}(px-q\xi)} dx\, d\xi.
\end{aligned}
$$

$$\tag{23.3}$$

Let

$$h = \frac{4c+1}{4\sqrt{r}}. \tag{23.4}$$

Then, by (23.3) and (23.4),

$$
\begin{aligned}
&(F *^{\frac{1}{2}} G)(z) \\
&= e^{(h^2-c)(q^2+p^2)} \int_{\mathbb{R}^2} e^{-(hq-\sqrt{r}x)^2} e^{-(hp-\sqrt{r}\xi)^2} e^{i(p\frac{x}{2}-q\frac{\xi}{2})} dx\, d\xi \\
&= 4e^{(h^2-c)(q^2+p^2)} \int_{\mathbb{R}^2} e^{-(hq-2\sqrt{r}x)^2} e^{-(hp-2\sqrt{r}\xi)^2} e^{i(px-q\xi)} dx\, d\xi
\end{aligned}
$$

$$= \quad 4e^{(h^2-c)(q^2+p^2)} \int_{\mathbb{R}^2} e^{-(hq-2\sqrt{r}x)^2} e^{-(hp-2\sqrt{r}\xi)^2}$$

$$[\cos(px - q\xi) + i\sin(px - q\xi)]dx\,d\xi$$

$$= \quad 4e^{(h^2-c)(q^2+p^2)} \int_{\mathbb{R}^2} e^{-(hq-2\sqrt{r}x)^2} e^{-(hp-2\sqrt{r}\xi)^2} \cos(px - q\xi)dx\,d\xi$$

$$+\, i4e^{(h^2-c)(q^2+p^2)} \int_{\mathbb{R}^2} e^{-(hq-2\sqrt{r}x)^2} e^{-(hp-2\sqrt{r}\xi)^2} \sin(px - q\xi)dx\,d\xi$$

$$= \quad 4e^{(h^2-c)(q^2+p^2)} I_1(z) + i4e^{(h^2-c)(q^2+p^2)} I_2(z) \qquad (23.5)$$

for all z in $\mathbb{C}$, where $I_1(z)$ and $I_2(z)$ are, respectively, the first integral and the second integral on the second last line of (23.5). Then

$$I_1(z) \quad = \quad \int_{\mathbb{R}^2} e^{-(hq-2\sqrt{r}x)^2} e^{-(hp-2\sqrt{r}\xi)^2} [\cos(px)\cos(q\xi) + \sin(px)\sin(q\xi)]dx\,d\xi$$

$$= \quad \int_{\mathbb{R}^2} e^{-(hq-2\sqrt{r}x)^2} e^{-(hp-2\sqrt{r}\xi)^2} \cos(px)\cos(q\xi)dx\,d\xi$$

$$+ \int_{\mathbb{R}^2} e^{-(hq-2\sqrt{r}x)^2} e^{-(hp-2\sqrt{r}\xi)^2} \sin(px)\sin(q\xi)dx\,d\xi$$

$$= \quad \int_{-\infty}^{\infty} e^{-(hq-2\sqrt{r}x)^2} \cos(px)dx \cdot \int_{-\infty}^{\infty} e^{-(hp-2\sqrt{r}\xi)^2} \cos(q\xi)d\xi$$

$$+ \int_{-\infty}^{\infty} e^{-(hq-2\sqrt{r}x)^2} \sin(px)dx \cdot \int_{-\infty}^{\infty} e^{-(hp-2\sqrt{r}\xi)^2} \sin(q\xi)d\xi \qquad (23.6)$$

for all z in $\mathbb{C}$. By Lemma 23.2,

$$\int_{-\infty}^{\infty} e^{-(hq-2\sqrt{r}x)^2} \cos(px)dx$$

$$= \frac{1}{2\sqrt{r}} \int_{-\infty}^{\infty} e^{-x^2} \cos\left[\frac{p}{2\sqrt{r}}(x + hq)\right] dx$$

$$= \frac{1}{2\sqrt{r}} \int_{-\infty}^{\infty} e^{-x^2} \left[\cos\left(\frac{p}{2\sqrt{r}}x\right)\cos\left(\frac{p}{2\sqrt{r}}hq\right) - \sin\left(\frac{p}{2\sqrt{r}}x\right)\sin\left(\frac{p}{2\sqrt{r}}hq\right)\right] dx$$

$$= \frac{1}{\sqrt{r}} \cos\left(\frac{p}{2\sqrt{r}}hq\right) \int_{0}^{\infty} e^{-x^2} \cos\left(\frac{p}{2\sqrt{r}}x\right) dx$$

$$= \frac{\sqrt{\pi}}{2\sqrt{r}} \cos\left(\frac{hqp}{2\sqrt{r}}\right) e^{-\frac{1}{16r}p^2} \qquad (23.7)$$

and

$$\int_{-\infty}^{\infty} e^{-(hq-2\sqrt{r}x)^2} \sin(px)dx$$

$$= \quad \frac{1}{2\sqrt{r}} \int_{-\infty}^{\infty} e^{-x^2} \sin\left[\frac{p}{2\sqrt{r}}(x + hq)\right] dx$$

$$= \frac{1}{2\sqrt{r}} \int_{-\infty}^{\infty} e^{-x^2} \left[\sin\left(\frac{p}{2\sqrt{r}}x\right) \cos\left(\frac{p}{2\sqrt{r}}hq\right) \right.$$

$$\left. + \cos\left(\frac{p}{2\sqrt{r}}x\right) \sin\left(\frac{p}{2\sqrt{r}}hq\right) \right] dx$$

$$= \frac{1}{\sqrt{r}} \sin\left(\frac{p}{2\sqrt{r}}hq\right) \int_0^{\infty} e^{-x^2} \cos\left(\frac{p}{2\sqrt{r}}x\right) dx$$

$$= \frac{\sqrt{\pi}}{2\sqrt{r}} \sin\left(\frac{hqp}{2\sqrt{r}}\right) e^{-\frac{1}{16r}p^2} \tag{23.8}$$

for all z in $\mathbb{C}$. So, by (23.7) and (23.8), (23.6) becomes

$$I_1(z) = \frac{\pi}{4r} \cos^2\left(\frac{hqp}{2\sqrt{r}}\right) e^{-\frac{1}{16r}(q^2+p^2)} + \frac{\pi}{4r} \sin^2\left(\frac{hqp}{2\sqrt{r}}\right) e^{-\frac{1}{16r}(q^2+p^2)}$$

$$= \frac{\pi}{4r} e^{-\frac{1}{16r}(q^2+p^2)} \tag{23.9}$$

for all z in $\mathbb{C}$. Similarly, by (23.7) and (23.8),

$$I_2(z) = \int_{\mathbb{R}^2} e^{-(hq-2\sqrt{r}x)^2-(hp-2\sqrt{r}\xi)^2} [\sin(px)\cos(q\xi) - \cos(px)\sin(q\xi)]dx\,d\xi$$

$$= \int_{\mathbb{R}^2} e^{-(hq-2\sqrt{r}x)^2-(hp-2\sqrt{r}\xi)^2} \sin(px)\cos(q\xi)dx\,d\xi$$

$$- \int_{\mathbb{R}^2} e^{-(hq-2\sqrt{r}x)^2-(hp-2\sqrt{r}\xi)^2} \cos(px)\sin(q\xi)dx\,d\xi$$

$$= \int_{-\infty}^{\infty} e^{-(hq-2\sqrt{r}x)^2} \sin(px)dx \cdot \int_{-\infty}^{\infty} e^{-(hp-2\sqrt{r}\xi)^2} \cos(q\xi)d\xi$$

$$- \int_{-\infty}^{\infty} e^{-(hq-2\sqrt{r}x)^2} \cos(px)dx \cdot \int_{-\infty}^{\infty} e^{-(hp-2\sqrt{r}\xi)^2} \sin(q\xi)d\xi$$

$$= \frac{\pi}{4r} \sin\left(\frac{hqp}{2\sqrt{r}}\right) \cos\left(\frac{hqp}{2\sqrt{r}}\right) e^{-\frac{1}{16r}(q^2+p^2)}$$

$$- \frac{\pi}{4r} \cos\left(\frac{hqp}{2\sqrt{r}}\right) \sin\left(\frac{hqp}{2\sqrt{r}}\right) e^{-\frac{1}{16r}(q^2+p^2)}$$

$$= 0 \tag{23.10}$$

for all z in $\mathbb{C}$. So, by (23.5), (23.9) and (23.10), we have

$$(F *^{\frac{1}{2}} G)(z) = \frac{\pi}{r} e^{(h^2-c-\frac{1}{16r})(q^2+p^2)} = \frac{\pi}{r} e^{(h^2-c-\frac{1}{16r})|z|^2}, \quad z \in \mathbb{C}.$$

Furthermore, by (23.2) and (23.4),

$$h^2 - c - \frac{1}{16r} = \frac{(4c+1)^2}{16r} - c - \frac{1}{16r} = \frac{c}{2r}(2c+1-2r)$$

$$= \frac{c}{2r}(2c+1-2c-2d-1) = -\frac{cd}{r}.$$

Therefore we get

$$(F *^{\frac{1}{2}} G)(z) = \frac{\pi}{r} e^{-\frac{cd}{r}|z|^2}, \quad z \in \mathbb{C}, \tag{23.11}$$

and the first step of the proof is complete. Next, setting $z = (z_1, z_2, \ldots, z_n)$ and $\zeta = (\zeta_1, \zeta_2, \ldots, \zeta_n)$, and using (23.11), we have

$$
\begin{aligned}
(F *^{\frac{1}{2}} G)(z) &= \int_{\mathbb{C}^n} e^{-c|z-\zeta|^2} e^{-d|\zeta|^2} e^{\frac{1}{2}(z\cdot\bar\zeta - |\zeta|^2)} d\zeta \\
&= \prod_{j=1}^{n} \int_{\mathbb{C}} e^{-c|z_j-\zeta_j|^2} e^{-d|\zeta_j|^2} e^{\frac{1}{2}(z_j\bar\zeta_j - |\zeta_j|^2)} d\zeta_j \\
&= \prod_{j=1}^{n} \frac{\pi}{r} e^{-\frac{cd}{r}|z_j|^2} = \left(\frac{\pi}{r}\right)^n e^{-\frac{cd}{r}|z|^2}
\end{aligned}
$$

for all z in $\mathbb{C}^n$, and the proof is complete. $\qquad\square$

The following result is also a consequence of the fact that the Fourier transform of the function ψ given by $\psi(x) = e^{-\frac{x^2}{2}}$, $x \in \mathbb{R}$, is equal to ψ.

Lemma 23.4 Let $\varphi(z) = e^{-t|z|^2}$, $t > 0$, $z \in \mathbb{C}^n$. Then $\hat\varphi(\zeta) = (2t)^{-n} e^{-\frac{1}{4t}|\zeta|^2}$, $\zeta \in \mathbb{C}^n$.

Now, we are ready to prove Theorem 23.1.

Proof of Theorem 23.1. By Theorem 22.2 and Corollary 22.3, it is enough to prove (23.1). By Lemma 23.4,

$$\hat{F}(\zeta) = \sum_{k=1}^{m} C_k \frac{1}{(2d_k)^n} e^{-\frac{1}{4d_k}|\zeta|^2}, \quad \zeta \in \mathbb{C}^n,$$

and

$$\hat{G}(\zeta) = \sum_{j=1}^{l} C'_j \frac{1}{(2d'_j)^n} e^{-\frac{1}{4d'_j}|\zeta|^2}, \quad \zeta \in \mathbb{C}^n.$$

So,

$$(\hat{F} *^{\frac{1}{2}} \hat{G})(\zeta) = \sum_{k=1}^{m} \sum_{j=1}^{l} C_k C'_j \frac{1}{(4d_k d'_j)^n} (F_k *^{\frac{1}{2}} G_j)(\zeta), \quad \zeta \in \mathbb{C}^n, \tag{23.12}$$

where

$$F_k(\zeta) = e^{-\frac{1}{4d_k}|\zeta|^2} = e^{-s_k|\zeta|^2}, \quad \zeta \in \mathbb{C}^n, \quad k = 1, 2, \ldots, m,$$

and

$$G_j(\zeta) = e^{-\frac{1}{4d'_j}|\zeta|^2} = e^{-t_j|\zeta|^2}, \quad \zeta \in \mathbb{C}^n, \quad j = 1, 2, \ldots, l,$$

where $s_k = \frac{1}{4d_k}$ and $t_j = \frac{1}{4d'_j}$. By Theorem 23.3 and (23.12),

$$(\hat{F} *^{\frac{1}{2}} \hat{G})(\zeta) = \sum_{k=1}^{m} \sum_{j=1}^{l} C_k C'_j \frac{1}{(4d_k d'_j)^n} \left(\frac{\pi}{h_{k,j}}\right)^n e^{-\frac{s_k t_j}{h_{k,j}} |\zeta|^2}, \quad \zeta \in \mathbb{C}^n,$$

where

$$h_{k,j} = s_k + t_j + \frac{1}{2} = \frac{1}{4d_k} + \frac{1}{4d'_j} + \frac{1}{2} = \frac{d'_j + d_k + 2d_k d'_j}{4d_k d'_j}.$$

Now,

$$\frac{s_k t_j}{h_{k,j}} = \frac{1}{16d_k d'_j} \frac{4d_k d'_j}{d'_j + d_k + 2d_k d'_j} = \frac{1}{4(d'_j + d_k + 2d_k d'_j)}.$$

Therefore, setting $r_{k,j} = d'_j + d_k + 2d_k d'_j$, we have

$$(\hat{F} *^{\frac{1}{2}} \hat{G})(\zeta) = \sum_{k=1}^{m} \sum_{j=1}^{l} C_k C'_j \left(\frac{\pi}{r_{k,j}}\right)^n e^{-\frac{1}{4r_{k,j}} |\zeta|^2}, \quad \zeta \in \mathbb{C}^n. \tag{23.13}$$

Then, by (22.4), (23.13) and Lemma 23.4 again, we get

$$\begin{aligned} H(z) &= \frac{1}{(2\pi)^n} \sum_{k=1}^{m} \sum_{j=1}^{l} C_k C'_j \left(\frac{\pi}{r_{k,j}}\right)^n (2r_{k,j})^n e^{-r_{k,j}|z|^2} \\ &= \sum_{k=1}^{m} \sum_{j=1}^{l} C_k C'_j e^{-r_{k,j}|z|^2} \end{aligned}$$

for all z in $\mathbb{C}^n$, and this completes the proof. $\qquad\qquad\square$

In fact, we can give a larger subspace of $L^2(\mathbb{C}^n)$ which contains the subspace M and has the same property as M with respect to the composition of Daubechies operators.

Let $\widetilde{M}$ be the set of all series of the form

$$\sum_{k=1}^{\infty} C_j e^{-d_j |z|^2}, \quad z \in \mathbb{C}^n,$$

where $C_1, C_2, \ldots$ are complex numbers, and $d_1, d_2, \ldots$ are positive numbers such that the sequences $\{C_j\}_{j=1}^{\infty}$ and $\left\{\frac{C_j}{d_j^n}\right\}_{j=1}^{\infty}$ are both in l^1. We call any such function a Gaussian series. It is clear that every Gaussian series is absolutely and uniformly convergent on $\mathbb{C}^n$.

Proposition 23.5 $\widetilde{M}$ *is a subspace of* $L^2(\mathbb{C}^n)$.

Proof. By Fubini's theorem, the assumption that the sequences $\{C_j\}_{j=1}^{\infty}$ and $\left\{\frac{C_j}{d_j^n}\right\}_{j=1}^{\infty}$ are both in l^1, and the fact that

$$\int_{\mathbb{C}^n} e^{-|z|^2} dz = \pi^n, \tag{23.14}$$

$$
\begin{aligned}
\left\| \sum_{j=1}^{\infty} C_j e^{-d_j |\cdot|^2} \right\|_{L^2(\mathbb{C}^n)}^2
&= \int_{\mathbb{C}^n} \left(\sum_{j=1}^{\infty} C_j e^{-d_j |z|^2} \right) \left(\sum_{k=1}^{\infty} \overline{C}_k e^{-d_k |z|^2} \right) dz \\
&= \int_{\mathbb{C}^n} \sum_{j=1}^{\infty} \sum_{k=1}^{\infty} C_j \overline{C}_k e^{-(d_j + d_k)|z|^2} dz \\
&= \sum_{j=1}^{\infty} \sum_{k=1}^{\infty} C_j \overline{C}_k \int_{\mathbb{C}^n} e^{-(d_j + d_k)|z|^2} dz \\
&= \sum_{j=1}^{\infty} \sum_{k=1}^{\infty} C_j \overline{C}_k \frac{\pi^n}{(d_j + d_k)^n} \\
&\leq \pi^n \sum_{j=1}^{\infty} |C_j| \sum_{k=1}^{\infty} \frac{|C_k|}{d_k^n} < \infty
\end{aligned}
$$

for all $\sum_{j=1}^{\infty} C_j e^{-d_j |\cdot|^2}$ in $\widetilde{M}$. Thus, every series in $\widetilde{M}$ is in $L^2(\mathbb{C}^n)$. To see that $\widetilde{M}$ is a subspace of $L^2(\mathbb{C}^n)$, let

$$\sum_{j=1}^{\infty} C_j e^{-d_j |\cdot|^2} \in \widetilde{M} \qquad \text{and} \qquad \sum_{j=1}^{\infty} C_j' e^{-d_j' |\cdot|^2} \in \widetilde{M},$$

and let $\alpha \in \mathbb{C}$. Then

$$\sum_{j=1}^{\infty} C_j e^{-d_j |z|^2} + \sum_{j=1}^{\infty} C_j' e^{-d_j' |z|^2} = \sum_{j=1}^{\infty} C_j'' e^{-d_j'' |z|^2}, \quad z \in \mathbb{C}^n,$$

where

$$C_{2j-1}'' = C_j, \quad d_{2j-1}'' = d_j \qquad \text{and} \qquad C_{2j}'' = C_j', \quad d_{2j}'' = d_j'$$

for $j = 1, 2, \ldots$. Thus,

$$
\begin{aligned}
\sum_{j=1}^{\infty} \frac{|C_j''|}{(d_j'')^n}
&= \sum_{j=1}^{\infty} \frac{|C_{2j-1}''|}{(d_{2j-1}'')^n} + \sum_{j=1}^{\infty} \frac{|C_{2j}''|}{(d_{2j}'')^n} \\
&= \sum_{j=1}^{\infty} \frac{|C_j|}{d_j^n} + \sum_{j=1}^{\infty} \frac{|C_j'|}{(d_j')^n} < \infty.
\end{aligned}
$$

Similarly, $\sum_{j=1}^{\infty} |C_j''| < \infty$. Hence

$$\sum_{j=1}^{\infty} C_j e^{-d_j|\cdot|^2} + \sum_{j=1}^{\infty} C_j' e^{-d_j'|\cdot|^2} \in \widetilde{M}.$$

That $\alpha \sum_{j=1}^{\infty} C_j e^{-d_j|\cdot|^2} \in \widetilde{M}$ is obvious. $\square$

We have the following result, which is better than Theorem 23.1.

Theorem 23.6 *Let F and G be in $\widetilde{M}$, i.e.,*

$$F(z) = \sum_{j=1}^{\infty} C_j e^{-d_j|z|^2}, \quad z \in \mathbb{C}^n,$$

and

$$G(z) = \sum_{k=1}^{\infty} C_k' e^{-d_k'|z|^2}, \quad z \in \mathbb{C}^n,$$

where C_1, C_2, ...; C_1', C_2', ... are complex numbers and d_1, d_2, ...; d_1', d_2', ... are positive numbers such that the sequences $\{C_j\}_{j=1}^{\infty}$, $\left\{\frac{C_j}{d_j^n}\right\}_{j=1}^{\infty}$, $\{C_k'\}_{k=1}^{\infty}$ and $\left\{\frac{C_k'}{d_k'^n}\right\}_{k=1}^{\infty}$ are all in l^1. Then $D_{F,\varphi}D_{G,\varphi} = D_{H,\varphi}$, where H is also in $\widetilde{M}$ and

$$H(z) = \sum_{j=1}^{\infty} \sum_{k=1}^{\infty} C_j C_k' e^{-r_{j,k}|z|^2}, \quad z \in \mathbb{C}^n, \tag{23.15}$$

where $r_{j,k} = d_j + d_k' + 2d_j d_k'$.

To prove this theorem, we need some preparation.

Lemma 23.7 *Let F be in $\widetilde{M}$, i.e.,*

$$F(z) = \sum_{j=1}^{\infty} C_j e^{-d_j|z|^2}, \quad z \in \mathbb{C}^n,$$

where C_1, C_2, ... are complex numbers, and d_1, d_2, ... are positive numbers such that the sequences $\{C_j\}_{j=1}^{\infty}$ and $\left\{\frac{C_j}{d_j^n}\right\}_{j=1}^{\infty}$ are both in l^1, and let F_m be the function on $\mathbb{C}^n$ given by

$$F_m(z) = \sum_{j=1}^{m} C_j e^{-d_j|z|^2}, \quad z \in \mathbb{C}^n.$$

Then

$$F_m \to F$$

in $L^2(\mathbb{C}^n)$ as $m \to \infty$.

Proof. By Fubini's theorem, the assumption that the sequences $\{C_j\}_{j=1}^{\infty}$ and $\left\{\frac{C_j}{d_j^n}\right\}_{j=1}^{\infty}$ are both in l^1, and (23.14),

$$
\begin{aligned}
\|F_m - F\|_{L^2(\mathbb{C}^n)}^2 &= \left\| \sum_{j=m+1}^{\infty} C_j e^{-d_j|\cdot|^2} \right\|_{L^2(\mathbb{C}^n)}^2 \\
&= \int_{\mathbb{C}^n} \left(\sum_{j=m+1}^{\infty} C_j e^{-d_j|z|^2} \right) \left(\sum_{k=m+1}^{\infty} \overline{C}_k e^{-d_k|z|^2} \right) dz \\
&= \int_{\mathbb{C}^n} \sum_{j=m+1}^{\infty} \sum_{k=m+1}^{\infty} C_j \overline{C}_k e^{-(d_j+d_k)|z|^2} dz \\
&= \sum_{j=m+1}^{\infty} \sum_{k=m+1}^{\infty} C_j \overline{C}_k \int_{\mathbb{C}^n} e^{-(d_j+d_k)|z|^2} dz \\
&= \sum_{j=m+1}^{\infty} \sum_{k=m+1}^{\infty} C_j \overline{C}_k \frac{\pi^n}{(d_j + d_k)^n} \\
&\leq \pi^n \sum_{j=m+1}^{\infty} |C_j| \sum_{k=m+1}^{\infty} \frac{|C_k|}{d_k^n} \to 0
\end{aligned}
$$

as $m \to \infty$. $\square$

Lemma 23.7 shows that, as subspaces of $L^2(\mathbb{C}^n)$, the subspace M spanned by the Gaussian functions is dense in $\widetilde{M}$.

Lemma 23.8 *Every F in $\widetilde{M}$ is in $L^1(\mathbb{C}^n)$.*

Proof. Let $F \in \widetilde{M}$. Then

$$
F(z) = \sum_{j=1}^{\infty} C_j e^{-d_j|z|^2}, \quad z \in \mathbb{C}^n,
$$

where $C_1, C_2, \ldots$ are complex numbers, and $d_1, d_2, \ldots$ are positive numbers such that the sequences $\{C_j\}_{j=1}^{\infty}$ and $\left\{\frac{C_j}{d_j^n}\right\}_{j=1}^{\infty}$ are in l^1. Then, by Fubini's theorem and (23.14),

$$
\begin{aligned}
\int_{\mathbb{C}^n} |F(z)| dz &\leq \int_{\mathbb{C}^n} \sum_{j=1}^{\infty} |C_j| e^{-d_j|z|^2} dz = \sum_{j=1}^{\infty} |C_j| \int_{\mathbb{C}^n} e^{-d_j|z|^2} dz \\
&= \sum_{j=1}^{\infty} \frac{|C_j|}{d_j^n} \int_{\mathbb{C}^n} e^{-|z|^2} dz = \pi^n \sum_{j=1}^{\infty} \frac{|C_j|}{d_j^n} < \infty.
\end{aligned}
$$

$\square$

Lemma 23.9 *Let F be in $\widetilde{M}$, i.e.,*

$$F(z) = \sum_{j=1}^{\infty} C_j e^{-d_j |z|^2}, \quad z \in \mathbb{C}^n,$$

where C_1, C_2, ... are complex numbers, and d_1, d_2, ... are positive numbers such that the sequences $\{C_j\}_{j=1}^{\infty}$ and $\left\{\frac{C_j}{d_j^n}\right\}_{j=1}^{\infty}$ are both in l^1. Then $\hat{F} \in \widetilde{M}$ and

$$\hat{F}(\zeta) = \sum_{j=1}^{\infty} C_j \frac{1}{(2d_j)^n} e^{-\frac{1}{4d_j}|\zeta|^2}, \quad \zeta \in \mathbb{C}^n.$$

Proof. It follows from Lemma 23.7 that

$$F_m = \sum_{k=1}^{m} C_j e^{-d_j |\cdot|^2} \to F$$

in $L^2(\mathbb{C}^n)$ as $m \to \infty$. So, by Plancherel's theorem and Lemma 23.4,

$$\hat{F}_m = \sum_{j=1}^{m} C_j \frac{1}{(2d_j)^n} e^{-\frac{1}{4d_j}|\cdot|^2} \to \hat{F} \tag{23.16}$$

in $L^2(\mathbb{C}^n)$ as $m \to \infty$. Therefore there exists a subsequence of $\{\hat{F}_m\}_{m=1}^{\infty}$, still denoted by $\{\hat{F}_m\}_{m=1}^{\infty}$, such that

$$\hat{F}_m(\zeta) \to \hat{F}(\zeta) \tag{23.17}$$

for almost all ζ in $\mathbb{C}^n$. Then, by (23.16) and (23.17),

$$\hat{F}(\zeta) = \sum_{j=1}^{\infty} C_j \frac{1}{(2d_j)^n} e^{-\frac{1}{4d_j}|\zeta|^2} \tag{23.18}$$

for almost all ζ in $\mathbb{C}^n$. By Lemma 23.8 and the Riemann-Lebesgue lemma, $\hat{F}$ is continuous on $\mathbb{C}^n$. Thus, (23.18) is true for all ζ in $\mathbb{C}^n$. That $\hat{F}$ is in $\widetilde{M}$ is now obvious.   □

Now, we are ready to prove Theorem 23.6.

Proof of Theorem 23.6. Let F and G be in $\widetilde{M}$. Then

$$F(z) = \sum_{j=1}^{\infty} C_j e^{-d_j |z|^2}, \quad z \in \mathbb{C}^n,$$

and

$$G(z) = \sum_{j=1}^{\infty} C'_j e^{-d'_j |z|^2}, \quad z \in \mathbb{C}^n,$$

where C_1, C_2, ...; C'_1, C'_2, ... are complex numbers and d_1, d_2, ...; d'_1, d'_2, ... are positive numbers such that the sequences

$$\{C_j\}_{j=1}^{\infty}, \quad \left\{\frac{C_j}{d_j^n}\right\}_{j=1}^{\infty}, \quad \{C'_k\}_{k=1}^{\infty} \quad \text{and} \quad \left\{\frac{C'_k}{d'^n_k}\right\}_{k=1}^{\infty}$$

are all in l^1. Then, by (23.12), Lemmas 23.8 and 23.9, Fubini's theorem and the Lebesgue dominated convergence theorem,

$$(\hat{F} *^{\frac{1}{2}} \hat{G})(\zeta)$$

$$= \int_{\mathbb{C}^n} \left(\sum_{j=1}^{\infty} \frac{C_j}{(2d_j)^n} e^{-\frac{1}{4d_j}|\zeta-\omega|^2}\right)\left(\sum_{j=1}^{\infty} \frac{C'_j}{(2d'_j)^n} e^{-\frac{1}{4d'_j}|\omega|^2}\right) e^{\frac{1}{2}(\zeta\cdot\bar{\omega}-|\omega|^2)} d\omega$$

$$= \int_{\mathbb{C}^n} \sum_{j=1}^{\infty}\sum_{k=1}^{\infty} \frac{C_j C'_k}{(4d_j d'_k)^n} e^{-\frac{1}{4d_j}|\zeta-\omega|^2} e^{-\frac{1}{4d'_k}|\omega|^2} e^{\frac{1}{2}(\zeta\cdot\bar{\omega}-|\omega|^2)} d\omega$$

$$= \sum_{j=1}^{\infty} \frac{C_j}{(2d_j)^n} \int_{\mathbb{C}^n} \lim_{l\to\infty} \sum_{k=1}^{l} \frac{C'_k}{(2d'_k)^n} e^{-\frac{1}{4d_j}|\zeta-\omega|^2} e^{-\frac{1}{4d'_k}|\omega|^2} e^{\frac{1}{2}(\zeta\cdot\bar{\omega}-|\omega|^2)} d\omega$$

$$= \sum_{j=1}^{\infty}\sum_{k=1}^{\infty} \frac{C_j C'_k}{(4d_j d'_k)^n} \int_{\mathbb{C}^n} e^{-\frac{1}{4d_j}|\zeta-\omega|^2} e^{-\frac{1}{4d'_k}|\omega|^2} e^{\frac{1}{2}(\zeta\cdot\bar{\omega}-|\omega|^2)} d\omega \tag{23.19}$$

for all ζ in $\mathbb{C}^n$. By (23.19) and the proof of Theorem 23.1,

$$(\hat{F} *^{\frac{1}{2}} \hat{G})(\zeta) = \sum_{j=1}^{\infty}\sum_{k=1}^{\infty} C_j C'_k \left(\frac{\pi}{r_{j,k}}\right)^n e^{-\frac{1}{4r_{j,k}}|\zeta|^2}, \quad \zeta \in \mathbb{C}^n. \tag{23.20}$$

Thus, by (23.20) and Proposition 23.5, $\hat{F} *^{\frac{1}{2}} \hat{G} \in L^2(\mathbb{C}^n)$. By Theorem 22.2, Corollary 22.3 and the fact that the function H defined by (23.15) is in $\widetilde{M}$, it is enough to prove that (22.4) is valid. But, by (23.15), Fubini's theorem and Lemma 23.4,

$$\hat{H}(\zeta) = (2\pi)^{-n} \int_{\mathbb{C}^n} e^{-iz\cdot\zeta} \sum_{j=1}^{\infty}\sum_{k=1}^{\infty} C_j C'_k e^{-r_{j,k}|z|^2} dz$$

$$= \sum_{j=1}^{\infty}\sum_{k=1}^{\infty} C_j C'_k (2\pi)^{-n} \int_{\mathbb{C}^n} e^{-iz\cdot\zeta} e^{-r_{j,k}|z|^2} dz$$

$$= \sum_{j=1}^{\infty}\sum_{k=1}^{\infty} C_j C'_k (2r_{j,k})^{-n} e^{-\frac{1}{4r_{j,k}}|\zeta|^2} \tag{23.21}$$

for all z in $\mathbb{C}^n$. So, by (23.20) and (23.21), $\hat{H} = (2\pi)^{-n}(\hat{F} *^{\frac{1}{2}} \hat{G})$ and the proof is complete. $\quad\square$

We give a remark on the commutativity of the product of Daubechies operators.

Remark 23.10 The product (or composition) of two Daubechies operators with symbols F and G in $\widetilde{M}$ is commutative because the $\frac{1}{2}$-convolution of $\hat{F}$ and $\hat{G}$ is commutative in view of (23.20). Thus, as explained in Section 1.1 in Chapter 1 of the book [4] by Berezin and Shubin, any finite collection of N Daubechies operators with symbols in $\widetilde{M}$ can be considered as a collection of N compatible quantum mechanical observables of which arbitrarily accurate simultaneous measurements can be made.

24   Group Actions and Homogeneous Spaces

A compact account of group actions and homogeneous spaces, which we give in this chapter, is helpful for a study of localization operators on homogeneous spaces in the next chapter. The book [2] by Ali, Antoine and Gazeau and the book [27] by Folland contain much more detailed material pertinent to this chapter.

Let Ω be a locally compact and Hausdorff topological space and let G be a locally compact and Hausdorff group. We say that G is a left transformation group on Ω if there exists a continuous mapping $G \times \Omega \ni (g, \omega) \mapsto g\omega \in \Omega$ such that for all g in G, the mapping $\Omega \ni \omega \mapsto g\omega \in \Omega$ is a homeomorphism of Ω onto Ω,

$$(gh)\omega = g(h\omega), \quad g, h \in G, \, \omega \in \Omega,$$

and

$$e\omega = \omega, \quad \omega \in \Omega,$$

where e is the identity element in G. The topological space Ω on which G acts is called a G–space and G is sometimes called a group action on Ω.

Let G be a left transformation group on Ω such that for all ω_1 and ω_2 in Ω, there exists an element g in G for which $\omega_2 = g\omega_1$. Then we say that the action of G on Ω is transitive and we call Ω a homogeneous space.

Proposition 24.1 *Let Ω be a homogeneous space on which G acts transitively. Let $\omega \in \Omega$. Then the set H_ω defined by*

$$H_\omega = \{g \in G : g\omega = \omega\}$$

is a closed subgroup of G.

Proof. H_ω is nonempty because $e \in H_\omega$. Let g and h be two elements in H_ω. Then $g\omega = \omega$ and $h\omega = \omega$. Thus,

$$(gh)\omega = g(h\omega) = g\omega = \omega.$$

So, $gh \in H_\omega$. Also,

$$\omega = e\omega = g^{-1}g\omega = g^{-1}\omega.$$

Thus, $g^{-1} \in H_\omega$. To see that H_ω is closed, let $g \notin H_\omega$. Then $g\omega \neq \omega$. Since Ω is Hausdorff, we can find a neighborhood U_1 of $g\omega$ and a neighborhood U_2 of ω such that $g\omega \in U_1$, $\omega \in U_2$ and $U_1 \cap U_2 = \phi$. Using the continuity of the mapping

$G \times \Omega \ni (g, \omega) \mapsto g\omega \in \Omega$, we can find a neighborhood N of g and a neighborhood U_3 of ω such that $N \times U_3 \subseteq U_1$. Thus, $h\omega \neq \omega$ for all h in N. Therefore $h \notin H_\omega$ for all h in N. This proves that the complement of H_ω in G is open and the proof is complete. $\qquad\qquad\qquad\qquad\qquad\qquad\qquad\qquad\qquad\qquad\qquad\qquad\qquad\qquad\quad\square$

We call H_ω the stability subgroup of G associated to ω.

Example 24.2 Let H be a closed subgroup of G and let $\Omega = G/H$, where

$$G/H = \{gH : g \in G\},$$

and gH, $g \in G$, is the left coset of g in H. Then the action of G on Ω defined by

$$G \times \Omega \ni (g, hH) \mapsto (gh)H \in \Omega, \quad g, h \in \Omega,$$

is transitive. Hence G/H is a homogeneous space.

Remark 24.3 All homogeneous spaces considered in this book are coset spaces given in Example 24.2.

Let Ω be a homogeneous space given by $\Omega = G/H$, where G is a locally compact and Hausdorff group and H is a closed subgroup of G. Let ν be a Borel measure on Ω. Then we say that ν is left invariant if

$$\nu(S) = \nu(gS), \quad g \in G,$$

for all Borel subsets S of Ω, where $gS = \{g\omega : \omega \in \Omega\}$. Thus, for all g in G,

$$\int_\Omega f(g\omega) d\nu(\omega) = \int_\Omega f(\omega) d\nu(g^{-1}\omega) = \int_\Omega f(\omega) d\nu(\omega)$$

for all Borel functions f on Ω. In view of Theorem 4.2, the group G always carries a left invariant measure, namely, the left Haar measure. Unfortunately, the homogeneous space G/H need not possess a left invariant Borel measure. A left quasi-invariant measure, though, does exist on G/H. A Borel measure ν on Ω is said to be left quasi-invariant if ν and ν_g are equivalent measures on Ω, where

$$\nu_g(S) = \nu(gS), \quad g \in G,$$

for all Borel subsets S of Ω.

We end this chapter with some notions that we need in the next chapter.

Let H be a closed subgroup of a locally compact and Hausdorff group G. Let $\Omega = G/H$. Then the mapping $q : G \to \Omega$ defined by

$$q(g) = gH, \quad g \in G,$$

is called the canonical surjection of G on Ω. A mapping $s : \Omega \to G$ is said to be a section on Ω if

$$q(s(\omega)) = \omega, \quad \omega \in \Omega.$$

25 A Unification

Localization operators in the setting of homogeneous spaces are first defined in this chapter. They are then shown to be in the trace class S_1 and a trace formula for them is given. Localization operators on locally compact and Hausdorff groups equipped with square-integrable representations, Daubechies operators and wavelet multipliers are then shown to be localization operators on homogeneous spaces. In this perspective, this chapter can be seen as a unification of the three classes of linear operators.

Let G be a locally compact and Hausdorff group and let H be a closed subgroup of G. Let ν be a left quasi-invariant measure on the homogeneous space $\Omega = G/H$. Let π be a unitary representation of G on a Hilbert space X. As usual, we denote the inner product and the norm in X by $(\,,\,)$ and $\|\,\|$ respectively. Let $s : \Omega \to G$ be a Borel section. Suppose that there exists an element φ in X such that $\|\varphi\| = 1$ and

$$\int_\Omega |(\varphi, \pi(s(\omega))\varphi)|^2 d\nu(\omega) < \infty.$$

Then we say that π is a square-integrable representation of G on X with respect to H and s, and we call φ an admissible wavelet for π. If φ is an admissible wavelet for the square-integrable representation π of G on X with respect to H and s, then we define the constant $c_{s,H,\varphi}$ by

$$c_{s,H,\varphi} = \int_\Omega |(\varphi, \pi(s(\omega))\varphi)|^2 d\nu(\omega).$$

Let $F \in L^1(\Omega)$. Then we define the linear operator $L_{F,s,H,\varphi} : X \to X$ by

$$(L_{F,s,H,\varphi}x, y) = \frac{1}{c_{s,H,\varphi}} \int_\Omega F(\omega)(x, \pi(s(\omega))\varphi)(\pi(s(\omega))\varphi, y) d\nu(\omega)$$

for all x and y in X.

Proposition 25.1 $L_{F,s,H,\varphi} : X \to X$ *is a bounded linear operator and*

$$\|L_{F,s,H,\varphi}\|_* \le \frac{1}{c_{s,H,\varphi}} \int_\Omega |F(\omega)| d\nu(\omega).$$

Proof. Using Schwarz' inequality, $\|\varphi\| = 1$ and the fact that $\pi(g) : X \to X$ is a unitary operator for all g in G, we get

$$|(L_{F,s,H,\varphi}x, y)| = \left| \frac{1}{c_{s,H,\varphi}} \int_\Omega F(\omega)(x, \pi(s(\omega))\varphi)(\pi(s(\omega))\varphi, y)d\nu(\omega) \right|$$

$$\leq \frac{1}{c_{s,H,\varphi}} \int_\Omega |F(\omega)| \, |(x, \pi(s(\omega))\varphi)| \, |(\pi(s(\omega))\varphi, y)|d\nu(\omega)$$

$$\leq \frac{1}{c_{s,H,\varphi}} \int_\Omega |F(\omega)|d\nu(\omega)\|x\|\|y\|$$

for all x and y in X, and the proof is complete.   □

Remark 25.2 If

$$(x, y) = \frac{1}{c_{s,H,\varphi}} \int_\Omega (x, \pi(s(\omega))\varphi)(\pi(s(\omega))\varphi, y)d\nu(\omega) \tag{25.1}$$

for all x and y in X, then the identity operator I on X can be written as

$$I = \frac{1}{c_{s,H,\varphi}} \int_\Omega (\cdot, \pi(s(\omega))\varphi)\pi(s(\omega))\varphi d\nu(\omega).$$

Thus, in this case, (25.1) plays the role of the resolution of the identity formula first formulated in (6.3). The role of the symbol $F : \Omega \to \mathbb{C}$ is again to localize on the homogeneous space Ω so as to produce a nontrivial bounded linear operator $L_{F,s,H,\varphi} : X \to X$ with applications in the mathematical sciences. So, we call the bounded linear operator $L_{F,s,H,\varphi} : X \to X$ a localization operator on the homogeneous space Ω.

We are now in a position to prove that localization operators on homogeneous spaces are in S_1 and compute their traces.

Proposition 25.3 *The localization operator* $L_{F,s,H,\varphi} : X \to X$ *is in* S_1.

Proof. As in the proof of Proposition 13.1, it is enough to prove the proposition for the case when F is a nonnegative and real-valued function on Ω. Let $\{\varphi_k : k = 1, 2, \ldots\}$ be an orthonormal basis for X. Then, using Fubini's theorem, Parseval's identity, $\|\varphi\| = 1$, and the fact that $\pi(g) : X \to X$ is a unitary operator for all g in G, we get

$$\sum_{k=1}^\infty (L_{F,s,H,\varphi}\varphi_k, \varphi_k) = \frac{1}{c_{s,H,\varphi}} \sum_{k=1}^\infty \int_\Omega F(\omega)|(\varphi_k, \pi(s(\omega))\varphi)|^2 d\nu(\omega)$$

$$\leq \frac{1}{c_{s,H,\varphi}} \int_\Omega |F(\omega)| \sum_{k=1}^\infty |(\varphi_k, \pi(s(\omega))\varphi)|^2 d\nu(\omega)$$

$$\leq \frac{1}{c_{s,H,\varphi}} \int_\Omega |F(\omega)|d\nu(\omega) < \infty, \tag{25.2}$$

and the proof is complete in view of Proposition 2.4.   □

Theorem 25.4 *The trace* $\mathrm{tr}(L_{F,s,H,\varphi})$ *of the localization operator* $L_{F,s,H,\varphi} : X \to X$ *is given by*

$$\mathrm{tr}(L_{F,s,H,\varphi}) = \frac{1}{c_{s,H,\varphi}} \int_{\Omega} F(\omega) d\nu(\omega).$$

Proof. Let $\{\varphi_k : k = 1, 2, \ldots\}$ be any orthonormal basis for X. Then, using Fubini's theorem, Parseval's identity, $\|\varphi\| = 1$ and the fact that $\pi(g) : X \to X$ is a unitary operator for all g in G, we get

$$
\begin{aligned}
\mathrm{tr}(L_{F,s,H,\varphi}) &= \sum_{k=1}^{\infty} (L_{F,s,H,\varphi}\varphi_k, \varphi_k) \\
&= \sum_{k=1}^{\infty} \frac{1}{c_{s,H,\varphi}} \int_{\Omega} F(\omega) |(\varphi_k, \pi(s(\omega))\varphi)|^2 d\nu(\omega) \\
&= \frac{1}{c_{s,H,\varphi}} \int_{\Omega} F(\omega) \sum_{k=1}^{\infty} |(\varphi_k, \pi(s(\omega))\varphi)|^2 d\nu(\omega) \\
&= \frac{1}{c_{s,H,\varphi}} \int_{\Omega} F(\omega) \|\pi(s(\omega))\|^2 d\nu(\omega) \\
&= \frac{1}{c_{s,H,\varphi}} \int_{\Omega} F(\omega) d\nu(\omega),
\end{aligned}
$$

and the proof is complete. $\qquad\qquad\qquad\qquad\qquad\qquad\qquad\qquad\qquad\quad\square$

We can now give three examples of localization operators on homogeneous spaces.

Example 25.5 (Locally Compact and Hausdorff Groups) Let $\varphi \in X$ be an admissible wavelet for a square-integrable representation $\pi : G \to U(X)$ of a locally compact and Hausdorff group G on X. Let $H = \{e\}$, where e is the identity element in the group G. Then, of course, $G/H = G$ and μ is a left invariant measure on the homogeneous space G/H, which is the same as G in this situation. The section $s : G/H \to G$ can be taken to be the identity mapping on G. Thus, the localization operator $L_{F,\varphi} : X \to X$ defined by (12.1) can be considered as a localization operator $L_{F,s,H,\varphi} : X \to X$ on the homogeneous space G/H.

Example 25.6 (Daubechies Operators) Let H be the subgroup of the Weyl-Heisenberg group $(WH)^n$ defined by

$$H = \{(0, 0, t) : t \in \mathbb{R}/2\pi\mathbb{Z}\}.$$

Then, of course, H is the center of $(WH)^n$, and the homogeneous space $(WH)^n/H$ is simply the Euclidean space $\mathbb{R}^n \times \mathbb{R}^n$. The Lebesgue measure on $\mathbb{R}^n \times \mathbb{R}^n$ is a left Haar measure on the homogeneous space $(WH)^n/H = \mathbb{R}^n \times \mathbb{R}^n$. Let $s : \mathbb{R}^n \times \mathbb{R}^n \to (WH)^n$ be the section defined by

$$s(q, p) = (q, p, 0), \quad (q, p) \in \mathbb{R}^n \times \mathbb{R}^n.$$

Then for all functions φ in $L^2(\mathbb{R}^n)$ with $\|\varphi\|_{L^2(\mathbb{R}^n)} = 1$, we get, by (17.19),

$$\int_{\mathbb{R}^n \times \mathbb{R}^n} |(\varphi, \pi(s(q,p))\varphi)_{L^2(\mathbb{R}^n)}|^2 dq\, dp = (2\pi)^{-n}$$

for all q and p in $\mathbb{R}^n$. Thus, every function φ in $L^2(\mathbb{R}^n)$ with $\|\varphi\|_{L^2(\mathbb{R}^n)} = 1$ is an admissible wavelet for the square-integrable representation π of $(WH)^n$ on $L^2(\mathbb{R}^n)$ with respect to H and s, and

$$c_{s,H,\varphi} = (2\pi)^{-n}.$$

Let $F \in L^1(\mathbb{R}^n \times \mathbb{R}^n)$. Then the localization operator $L_{F,s,H,\varphi} : L^2(\mathbb{R}^n) \to L^2(\mathbb{R}^n)$ on the homogeneous space $(WH)^n/H$ is exactly the same as the Daubechies operator $D_{F,\varphi} : L^2(\mathbb{R}^n) \to L^2(\mathbb{R}^n)$ defined by (17.23).

Example 25.7 To see that the wavelet multipliers introduced in Chapter 19 are in fact localization operators on a homogeneous space, we let H be the subgroup of $\mathbb{R}^n$ given by $H = \{0\}$, where 0 is the additive identity of the group $\mathbb{R}^n$. Then, obviously, $\mathbb{R}^n/H = \mathbb{R}^n$ and the Lebesgue measure on $\mathbb{R}^n$ is a left invariant measure on the homogeneous space $\mathbb{R}^n/H$, which is simply the same as $\mathbb{R}^n$. The section $s : \mathbb{R}^n/H \to \mathbb{R}^n$ is taken to be the identity mapping on $\mathbb{R}^n$. Let $\varphi \in L^2(\mathbb{R}^n) \cap L^4(\mathbb{R}^n)$ be such that $\|\varphi\|_{L^2(\mathbb{R}^n)} = 1$. Then

$$
\begin{aligned}
c_{s,H,\varphi} &= \int_{\mathbb{R}^n} |(\varphi, \pi(s(\xi)\varphi)_{L^2(\mathbb{R}^n)}|^2 d\xi = \int_{\mathbb{R}^n} |(\varphi, \pi(\xi)\varphi)_{L^2(\mathbb{R}^n)}|^2 d\xi \\
&= \int_{\mathbb{R}^n} \left| \int_{\mathbb{R}^n} e^{ix\cdot\xi} |\varphi(x)|^2 dx \right|^2 d\xi = (2\pi)^n \|\varphi\|_{L^4(\mathbb{R}^n)}^4.
\end{aligned}
$$

Thus, every function φ in $L^2(\mathbb{R}^n) \cap L^4(\mathbb{R}^n)$ with $\|\varphi\|_{L^2(\mathbb{R}^n)} = 1$ is an admissible wavelet for the unitary representation of $\mathbb{R}^n$ on $L^2(\mathbb{R}^n)$ with respect to H and s. Furthermore, let $\sigma \in L^1(\mathbb{R}^n)$. Then

$$
\begin{aligned}
&(L_{\sigma,s,H,\varphi} u, v)_{L^2(\mathbb{R}^n)} \\
={}& \frac{1}{c_{s,H,\varphi}} \int_{\mathbb{R}^n} \sigma(\xi)(u, \pi(\xi)\varphi)_{L^2(\mathbb{R}^n)}(\pi(\xi)\varphi, v)_{L^2(\mathbb{R}^n)} d\xi \\
={}& (2\pi)^{-n} \|\varphi\|_{L^4(\mathbb{R}^n)}^{-4} \int_{\mathbb{R}^n} \sigma(\xi)(u, \pi(\xi)\varphi)_{L^2(\mathbb{R}^n)}(\pi(\xi)\varphi, v)_{L^2(\mathbb{R}^n)} d\xi \\
={}& \|\varphi\|_{L^4(\mathbb{R}^n)}^{-4} (P_{\sigma,\varphi} u, v)_{L^2(\mathbb{R}^n)}
\end{aligned}
$$

for all u and v in $L^2(\mathbb{R}^n)$. Thus, the localization operator $L_{\sigma,s,H,\varphi} : L^2(\mathbb{R}^n) \to L^2(\mathbb{R}^n)$ on the homogeneous space $\mathbb{R}^n/H$ is a scalar multiple of the wavelet multiplier $P_{\sigma,\varphi} : L^2(\mathbb{R}^n) \to L^2(\mathbb{R}^n)$.

26 The Affine Group Action on $\mathbb{R}$

The aim of this chapter is to show that localization operators on $\mathbb{R}$ considered as a homogeneous space under the action of the affine group U are wavelet multipliers.

First, we look at the mapping $U \times \mathbb{R} \ni ((b,a), x) \mapsto ax + b \in \mathbb{R}$. It is obvious that U acts on $\mathbb{R}$ transitively. In other words, $\mathbb{R}$ is a homogeneous space on which the transitive group action is given by the affine group U. Let H be the closed subgroup of U given by

$$H = \{(0, a) : a > 0\}.$$

Proposition 26.1 *The homogeneous space U/H is isomorphic to $\mathbb{R}$ as topological groups.*

Proof. For all $(b, a) \in U$, we get

$$(b, a) = (b, 1) \cdot (0, a).$$

Thus,

$$(b, a)H = (b, 1)H, \quad (b, a) \in U.$$

Hence the mapping $\mathbb{R} \ni b \mapsto (b, 1)H \in U/H$ is bijective. That the mapping is a homeomorphism is obvious. Finally, for all b_1 and b_2 in $\mathbb{R}$, we get

$$(b_1, 1)H \cdot (b_2, 1)H = ((b_1, 1) \cdot (b_2, 1))H = (b_1 + b_2, 1)H,$$

and hence the mapping is also a group homomorphism. $\qquad\qquad\square$

Proposition 26.2 *The Lebesgue measure on $\mathbb{R}$ is a left quasi-invariant measure on $\mathbb{R}$ considered as a homogeneous space under the group action U.*

Proof. Let ν be the Lebesgue measure on $\mathbb{R}$. For all (b, a) in U, we have

$$d\nu_{(b,a)}(x) = d\nu(ax + b) = a\,d\nu(x),$$

and hence the measures ν and $\nu_{(b,a)}$ are equivalent. $\qquad\qquad\square$

Let $s : \mathbb{R} \to U$ be the global section, i.e., $s(x) = (x, 1)$, $x \in \mathbb{R}$. Let π be the unitary representation of U on $L^2(\mathbb{R})$ defined by

$$(\pi(b, a)u)(x) = \frac{1}{\sqrt{a}} u\left(\frac{x - b}{a}\right), \quad x \in \mathbb{R},$$

for all (b,a) in U and all u in $L^2(\mathbb{R})$. Let φ be any function in $L^2(\mathbb{R})$ such that $\|\varphi\|_{L^2(\mathbb{R})} = 1$ and $\hat{\varphi} \in L^4(\mathbb{R})$. Then, using Plancherel's theorem, we get

$$
\begin{aligned}
c_{s,H,\varphi} &= \int_{-\infty}^{\infty} |(\varphi, \pi(s(x))\varphi)_{L^2(\mathbb{R})}|^2 dx \\
&= \int_{-\infty}^{\infty} |(\varphi, \pi(x,1)\varphi)_{L^2(\mathbb{R})}|^2 dx \\
&= \int_{-\infty}^{\infty} |(\varphi, T_{-x}\varphi)_{L^2(\mathbb{R})}|^2 dx \\
&= \int_{-\infty}^{\infty} \left| \int_{-\infty}^{\infty} e^{ix\xi} |\hat{\varphi}(\xi)|^2 d\xi \right|^2 dx \\
&= 2\pi \|\hat{\varphi}\|_{L^4(\mathbb{R})}^4,
\end{aligned}
$$

where $(T_{-x}\varphi)(b) = \varphi(b-x)$, $b \in \mathbb{R}$.

Let $\sigma \in L^1(\mathbb{R})$. The localization operator $L_{\sigma,s,H,\varphi} : L^2(\mathbb{R}) \to L^2(\mathbb{R})$ on $\mathbb{R}$, considered as a homogeneous space under the group action U, is defined in Chapter 25. In order to understand the localization operator $L_{\sigma,s,H,\varphi} : L^2(\mathbb{R}) \to L^2(\mathbb{R})$ better, we suppose that σ is also in $L^\infty(\mathbb{R})$. Then for all u and v in $C_0(\mathbb{R})$, we can use Plancherel's theorem and a change of variables to obtain

$$
(u, \pi(s(x))\varphi)_{L^2(\mathbb{R})} = (u, T_{-x}\varphi)_{L^2(\mathbb{R})} = \int_{-\infty}^{\infty} e^{ix\xi} \hat{u}(\xi) \overline{\hat{\varphi}(\xi)} d\xi = (2\pi)^{\frac{1}{2}} \widehat{(\hat{u}\overline{\hat{\varphi}})}(-x)
$$

for all x in $\mathbb{R}$. Thus, for all u and v in $L^2(\mathbb{R})$, we get

$$
\begin{aligned}
&(L_{\sigma,s,H,\varphi} u, v)_{L^2(\mathbb{R})} \\
&= \|\hat{\varphi}\|_{L^4(\mathbb{R})}^{-4} \int_{-\infty}^{\infty} \tilde{\sigma}(x) \widehat{(\hat{u}\overline{\hat{\varphi}})}(x) \overline{\widehat{(\hat{v}\overline{\hat{\varphi}})}(x)} dx \\
&= \|\hat{\varphi}\|_{L^4(\mathbb{R})}^{-4} \int_{-\infty}^{\infty} (T_{\tilde{\sigma}}(\overline{\hat{\varphi}}\hat{u}))(x) \overline{(\overline{\hat{\varphi}}\hat{v})(x)} dx \\
&= \|\hat{\varphi}\|_{L^4(\mathbb{R})}^{-4} (T_{\tilde{\sigma}}(\overline{\hat{\varphi}}\hat{u}), \overline{\hat{\varphi}}\hat{v})_{L^2(\mathbb{R})} \\
&= \|\hat{\varphi}\|_{L^4(\mathbb{R})}^{-4} (\hat{\varphi} T_{\tilde{\sigma}}\overline{\hat{\varphi}}\hat{u}, \hat{v})_{L^2(\mathbb{R})} \\
&= \|\hat{\varphi}\|_{L^4(\mathbb{R})}^{-4} (\mathcal{F}^{-1}(\hat{\varphi} T_{\tilde{\sigma}}\overline{\hat{\varphi}})\mathcal{F}u, v)_{L^2(\mathbb{R})},
\end{aligned}
$$

where $\tilde{\sigma}(x) = \sigma(-x)$, $x \in \mathbb{R}$.

Therefore the localization operator $L_{\sigma,s,H,\varphi} : L^2(\mathbb{R}) \to L^2(\mathbb{R})$ is unitarily equivalent to the linear operator $\|\hat{\varphi}\|_4^{-4} P_{\sigma,\hat{\varphi}} : L^2(\mathbb{R}) \to L^2(\mathbb{R})$, where $P_{\sigma,\hat{\varphi}} : L^2(\mathbb{R}) \to L^2(\mathbb{R})$ is the wavelet multiplier associated to the symbol σ and the admissible wavelet $\hat{\varphi}$ studied in Chapter 19.

References

[1] I. J. R. Aitchison, Relativistic Quantum Mechanics, Macmillan, 1972.

[2] S. T. Ali, J.-P. Antoine and J.-P. Gazeau, Coherent States, Wavelets and Their Generalizations, Springer-Verlag, 2000.

[3] S. T. Ali, J.-P.Antoine, J.-P. Gazeau and U. A. Mueller, Coherent states and their generalizations: A mathematical overview, Rev. Math. Phys. 7 (1995), 1013–1104.

[4] F. A. Berezin and M. A. Shubin, The Schrödinger Equation, Kluwer Academic Publishers, 1991.

[5] J. Bergh and J. Löfström, Interpolation Spaces, An Introduction, Springer-Verlag, 1976.

[6] J. D. Bjorken and S. D. Drell, Relativistic Quantum Mechanics, McGraw-Hill, 1964.

[7] C. Blatter, Wavelets: A Primer, A. K. Peters, 1998

[8] P. Boggiatto, E. Buzano and L. Rodino, Global Hypoellipticity and Spectral Theory, Akademie-Verlag, 1996.

[9] A. L. Carey, Square integrable representations of non-unimodular groups, Bull. Austral. Math. Soc. 15 (1976), 1–12.

[10] Y. T. Chan, Wavelet Basics, Kluwer Academic Publishers, 1995.

[11] C. K. Chui, An Introduction to Wavelets, Academic Press, 1992.

[12] I. Daubechies, Time-frequency localization operators: a geometric phase space approach, IEEE Trans. Inform. Theory 34 (1988), 605–612.

[13] I. Daubechies, Ten Lectures on Wavelets, SIAM, 1992.

[14] J. Dixmier, C^*-Algebras, North-Holland, 1977

[15] R. G. Douglas, Banach Algebra Techniques in Operator Theory, Second Edition, Springer-Verlag, 1998.

[16] J. Du and M. W. Wong, A trace formula for Weyl transforms with radial symbols, Integral Equations Operator Theory 37 (2000), 232–237.

[17] J. Du and M. W. Wong, A trace formula for Weyl transforms, Approx. Theory Applic. 16 (2000), 41–45.

[18] J. Du and M. W. Wong, A product formula for localization operators, Bull. Korean Math. Soc. 37 (2000), 77–84.

[19] J. Du and M. W. Wong, Traces of localization operators, C. R. Math. Rep. Acad. Sci. Canada 22 (2000), 92–95.

[20] J. Du and M. W. Wong, Gaussian functions and Daubechies operators, Integral Equations Operator Theory 38 (2000), 1–8.

[21] J. Du and M. W. Wong, Gaussian Series and Daubechies operators, Applic. Anal. 76(1–2) (2000), 83–91.

[22] J. Du and M. W. Wong, Traces of wavelet multipliers, C. R. Math. Rep. Acad. Sci. Canada 23 (2001), 148–152.

[23] J. Du, M. W. Wong and Z. Zhang, Trace class norm inequalities for localization operators, Integral Equations Operator Theory 41 (2001), 497–503.

[24] M. Duflo and C. C. Moore, On the regular representation of a non-unimodular locally compact group, J. Funct. Anal. 21 (1976), 209–243.

[25] N. Dunford and J. T. Schwartz, Linear Operators, Part II, Wiley, 1963.

[26] G. B. Folland, Harmonic Analysis in Phase Space, Princeton University Press, 1989.

[27] G. B. Folland, A Course in Abstract Harmonic Analysis, CRC Press, 1995

[28] D. Gabor, Theory of communications, J. Inst. Elec. Eng. (London) 93 (1946), 429–457.

[29] C. Gasquet and P. Witomsky, Fourier Analysis and Applications: Filtering, Numerical Computation, Wavelets, Springer-Verlag, 1999.

[30] I. Gohberg and S. Goldberg, Basic Operator Theory, Birkhäuser, 1981.

[31] I. Gohberg, S. Goldberg and N. Krupnik, Traces and Determinants of Linear Operators, Birkhäuser, 2000.

[32] R. R. Goldberg, Fourier Transforms, Cambridge University Press, 1961.

[33] K. Gröchenig, Foundations of Time-Frequency Analysis, Birkhäuser, 2001.

[34] K. Gröchenig and C. Heil, Modulation spaces and pseudodifferential operators, Integral Equations Operator Theory, 34 (1999), 439–457.

[35] A. Grossmann, G. Loupias and E. M. Stein, An algebra of pseudodifferential operators and quantum mechanics in phase space, Ann. Inst. Fourier (Grenoble) 18 (1968), 343–368.

[36] A. Grossmann, J. Morlet and T. Paul, Transforms associated to square integrable group representations I: General results, J. Math. Phys. 26 (1985), 2473–2479.

[37] J. He, Continuous multiscale analysis on the Heisenberg group, Bull. Korean Math. Soc. 38 (2001), 517–216.

[38] J. He and H. Liu, Admissible wavelets associated with the affine automorphism group of the Siegel upper half plane, J. Math. Anal. Appl. 208 (1997), 58–70.

[39] Z. He, A unique continuation property for wavelet transforms, Panamer. Math. J. 8(1) (1998), 65–78.

[40] Z. He and M. W. Wong, Localization operators associated to square integrable group representations, Panamer. Math. J. 6 (1) (1996), 93–104.

[41] Z. He and M. W. Wong, Wavelet multipliers and signals, J. Austral. Math. Soc. Ser. B 40 (1999), 437–446.

[42] C. Heil, J. Ramanathan and P. Topiwala, Singular values of compact pseudodifferential operators, J. Funct. Anal. 150 (1997), 426–452.

[43] C. Heil and D. Walnut, Continuous and discrete wavelet transforms, SIAM Rev. 31(1989), 628–666.

[44] E. Hernández and G. Weiss, A First Course on Wavelets, CRC Press, 1996.

[45] I. N. Herstein, Topics in Algebra, Second Edition, Xerox, 1975.

[46] M. Holschneider, Wavelets: An Analysis Tool, Oxford University Press, 1995.

[47] L. Hörmander, The Analysis of Linear Partial Differential Operators III, Springer-Verlag, 1985.

[48] Q. Jiang and L. Peng, Wavelet transform and Toeplitz-Hankel type operators, Math. Scand. 70 (1992), 247–264.

[49] G. Kaiser, A Friendly Guide to Wavelets, Birkhäuser, 1994.

[50] T. Kato, Perturbation Theory for Linear Operators, Second Edition, Springer-Verlag, 1976.

[51] T. Kawazoe, Wavelet transforms associated to a principal series representation of semisimple Lie groups I, Proc. Japan Acad. 71 (1995), 154–157.

[52] T. Kawazoe, Wavelet transforms associated to a principal series representation of semisimple Lie groups II, Proc. Japan Acad. 71 (1995), 158–160.

[53] T. Kawazoe, Wavelet transform associated to an induced representation of $SL(n + 2, \mathbb{R})$, Ann. Inst. Henri Poincaré 65 (1996), 1–13.

[54] J. L. Kelley, General Topology, Springer-Verlag, 1997.

[55] A. A. Kirillov, Elements of the Theory of Representations, Springer-Verlag, 1976.

[56] J. R. Klauder and B.-S. Skagerstam, Coherent States, World Scientific, 1985.

[57] S. G. Krantz, A Panorama of Harmonic Analysis, The Mathematical Association of America, 1999.

[58] H. Kumano-go, Pseudo-Differential Operators, MIT Press, 1981.

[59] H. J. Landau and H. O. Pollak, Prolate spheroidal wave functions, Fourier analysis and uncertainty, II, Bell Syst. Tech. J. 40 (1961), 65–84.

[60] H. J. Landau and H. O. Pollak, Prolate spheroidal wave functions, Fourier analysis and uncertainty, III, Bell Syst. Tech. J. 41 (1962), 1295–1336.

[61] H. Liu, Wavelet transforms and symmetric tube domains, J. Lie Theory 8 (1998), 351–366.

[62] H. Liu and L. Peng, Admissible wavelets associated with the Heisenberg group, Pacific J. Math. 180 (1997), 101–123.

[63] Y. Meyer, Wavelets: Algorithms and Applications, SIAM, 1993.

[64] J. R. Munkres, Topology: A First Course, Second Edition, Upper Saddle River: Prentice-Hall, 2000.

[65] R. Peierls, On a minimization property of the free energy, Phys. Rev. 54 (1938), 918–919.

[66] A. Perelomov, Generalized Coherent States and Their Applications, Springer-Verlag, 1986.

[67] M. A. Pinsky, Introduction to Fourier Analysis and Wavelets, Brooks/Cole, 2002.

[68] L. Pontragin, Topological Groups, Princeton University Press, 1939.

[69] J. C. T. Pool, Mathematical aspects of the Weyl correspondence, J. Math. Phys. 7 (1966), 66–76.

[70] J. Ramanathan and P. Topiwala, Time-frequency localization via the Weyl correspondence, SIAM J. Math. Anal. 24 (1993), 1378–1393.

[71] R. M. Rao and A. S. Bopardikar, Wavelet Transforms: Introduction to Theory and Applications, Addison-Wesley, 1998.

[72] M. Reed and B. Simon, Methods of Modern Mathematical Physis I: Functional Analysis, Revised and Enlarged Edition, Academic Press, 1980.

[73] H. L. Royden, Real Analysis, Third Edition, Prentice-Hall, 1988.

[74] W. Rudin, Real and Complex Analysis, Third Edition, McGraw-Hill, 1988.

[75] X. Saint Raymond, Elementary Introduction to the Theory of Pseudodifferential Operators, CRC Press, 1991.

[76] M. Schechter, Principles of Functional Analysis, Academic Press, 1971.

[77] M. Schechter, Spectra of Partial Differential Operators, Second Edition, North-Holland, 1986.

[78] E. Schrödinger, Der stetige Übergang von der Mikro- zur Makromechanik, Naturwiss. 14 (1926), 664–666.

[79] C. E. Shannon, A mathematical theory of communication, Bell Syst. Tech. J. 27 (1948), 379–423.

[80] C. E. Shannon, A mathematical theory of communication, Bell. Syst. Tech. J. 27 (1948), 623–656.

[81] M. A. Shubin, Pseudodifferential Operators and Spectral Theory, Springer-Verlag, 1987.

[82] B. Simon, Trace Ideals and Their Applications, Cambridge University Press, 1979.

[83] D. Slepian, On bandwidth, Proc. IEEE 64 (1976), 292–300.

[84] D. Slepian, Some comments on Fourier analysis, uncertainty and modeling, SIAM Rev. 25 (1983), 379–393.

[85] D. Slepian and H. O. Pollak, Prolate spheroidal wave functions, Fourier analysis and uncertainty, I, Bell Syst. Tech. J. 40 (1961), 43–64.

[86] H. G. Stark, Continuous wavelet transform and continuous multiscale analysis, J. Math. Anal. Appl. 169 (1992), 179–196.

[87] E. M. Stein, Harmonic Analysis: Real-Variable Methods, Orthogonality and Oscillatory Integrals, Princeton University Press, 1993.

[88] E. M. Stein and G. Weiss, Introduction to Fourier Analysis on Euclidean Spaces, Princeton University Press, 1971.

[89] G. Strang and T. Nguyen, Wavelets and Filter Banks, Revised Edition, Wellesley-Cambridge Press, 1997.

[90] R. S. Strichartz, How to make wavelets, Amer. Math. Monthly 100 (1993), 539–556.

[91] K. Symanzik, Proof and refinements of an inequality of Feynman, J. Math. Phys. 6 (1965), 1155–1156

[92] K. Tachizawa, The boundedness of pseudodifferential operators on modulation spaces, Math. Nachr. 168 (1994), 263–277.

[93] M. E. Taylor, Pseudodifferential Operators, Princeton University Press, 1981.

[94] S. Thangavelu, Lectures on Hermite and Laguerre Expansions, Princeton University Press, 1993.

[95] S. Thangavelu, Harmonic Analysis on the Heisenberg Group, Birkhäuser, 1998.

[96] F. Treves, Introduction to Pseudodifferential and Fourier Integral Operators, Volume 1: Pseudodifferential Operators, Plenum Press, 1980.

[97] J. von Neumann, Mathematical Foundations of Quantum Mechanics, Princeton University Press, 1996.

[98] J. S. Walker, Fourier analysis and wavelet analysis, Notices Amer. Math. Soc. 44 (1997), 658–670.

[99] J. S. Walker, A Primer on Wavelets and Their Scientific Applications, Chapman and Hall/CRC, 1999.

[100] H. Weyl, The Theory of Groups and Quantum Mechanics, Dover, 1950.

[101] P. Wojtaszczyk, A Mathematical Introduction to Wavelets, Cambridge University Press, 1997.

[102] M. W. Wong, Weyl Transforms, Springer-Verlag, 1998.

[103] M. W. Wong, An Introduction to Pseudo-Differential Operators, Second Edition, World Scientific, 1999.

[104] M. W. Wong, Localization Operators, Lecture Notes Series 47, Seoul National University, Research Institute of Mathematics, Global Analysis Research Center, Seoul, 1999.

[105] M. W. Wong, Localization operators on the Weyl-Heisenberg group, in Geometry, Analysis and Applications, Editor: R. S. Pathak, World Scientific, 2001, 303–314.

[106] N. Young, An Introduction to Hilbert Space, Cambridge University Press, 1988.

[107] R. K. Young, Wavelet Theory and Its Applications, Kluwer Academic Publishers, 1993.

[108] K. Zhu, Operator Theory in Function Spaces, Marcel Dekker, 1990.

Index